MONTPELIER REGIONAL LIBRARY
RFD #2
MONTPELIER, VT. 05602

WITHDRAWN

Beginnings and Blunders

BOOKS BY LANCELOT HOGBEN

Mathematics for the Million
Science for the Citizen
Mathematics in the Making
From Cave Painting to Comic Strip
The Mother Tongue

LANCELOT HOGBEN

Beginnings and Blunders

or Before Science Began

A W.W. NORTON BOOK

Published by Grosset & Dunlap Inc. New York

Copyright © 1970 by Lancelot Hogben
Grosset & Dunlap Edition 1971

Library of Congress Catalogue Card No : 73-119516

ISBN: 0-448-21400-8 (Trade Edition)
ISBN: 0-448-26174-X (Library Edition)

All rights reserved

Published simultaneously in Canada

Printed in the United States of America

Contents

Acknowledgements	vi
Foreword	xi
Chapter 1 The Age of our Earth and our Ancestors	1
Chapter 2 What our Ancestors were like	16
Chapter 3 Food, Fires and Fur Coats	27
Chapter 4 The Domestication of Man	43
Chapter 5 Crops and Containers	55
Chapter 6 Fellers of Trees, Fishermen, Trappers and Traders	67
Chapter 7 Handicrafts among the First Farmers	78
Chapter 8 Cradles of Civilization	92

Acknowledgements

The author is indebted to the following for permission to reproduce pictures on the pages indicated: George Allen & Unwin Ltd, 4, 6, 12, 28, 36, 37, 40, 41, 60, 65, 69, 80, 81, 84, 87, 96; The American Museum of Natural History, 89; BPC Publishing Ltd, 26, 32, 33, 34, 35, 44; The Trustees of the British Museum, 9, 11, 17, 18, 19, 21, 24, 29, 39, 40, 66, 74, 99, 104–5; The Trustees of the British Museum (Natural History), 7, 9; The Cambridge University Museum of Archaeology and Ethnology, 82; The Cambridge University Press, 2; Jonathan Cape Ltd, 3, 27, 47; Professor Grahame Clark, 71; Mrs Sonia Cole, 84; Miss Dorothy Davison, 3; *Endeavour*, 86; Mexican National Tourist Council, 109; The Oriental Institute of the University of Chicago, 79; Professor Doctor Elisabeth Schiemann (for drawing from her book *Entstehung der Kulturpflanzen*), 56, 57; The Worthing Museum (for drawing from E. C. Curwen's *Archaeology of Sussex*), 77.

Illustrations

Chapter 1
1. *Homo diluvii testis*
2. Palaeolithic implements
    (a) Old World—
        (i) flint implement of Mousterian type
        (ii) Solutrean flint implements
    (b) New World—
        (a) Clovis point
        (b) Folsom point
3. Neolithic implements
    (a) double-edged axes and hammer axe of polished stone
    (b) mace-head and battle-axes
4. Australian saw-knife
5. Australian using hand-axe
6. Methods of making tools
7. Cave painting of bow and arrow hunt

Chapter 2
1. Palate of *Australopithecus* compared with man and gorilla
2. Pelvis of *Australopithecus* compared with man and gorilla
3. Skulls of chimpanzee, *Homo erectus* and *Homo sapiens*
4. The Neanderthal skull compared with a modern human skull
5. Rhodesian man
6. Early cave paintings

Chapter 3
1. Cave paintings of animals
2. Cave paintings of women's attire
3. Cave paintings of human behaviour
4a & 4b. Animal totem star clusters

5a, 5b & 5c   Animal disguises
6. Upper Palaeolithic fur clothing—ivory carving
7. Old Stone Age Venuses
8. Eskimo clothing
9. Bone needles for making clothing
10. Reed hut building
    (a) basic structure
    (b) completed building

Chapter 4
1. Bushman painting of rhinoceros hunt with dog
2. Dog buried in Egyptian Neolithic site

Chapter 5
1. Wild species of wheat
    (a) diploid wheat
    (b) tetraploid wheat
2. Palaeolithic stone lamps
3. Mesolithic and Neolithic sickles

Chapter 6
1. Old Stone Age harpoons
2. Mesolithic fishing gear
3. Stone Age rock picture of boat
4. Log canoe made by burning out a cavity
5. Antler picks and scapula shovels

Chapter 7
1. Neolithic house in Iraq
2. Basket-lined grain storage pit
3. Early basketry from Peru
4. Contemporary basket making
5. Pottery jar from Kenya
6. Neolithic Egyptian pottery

7. Primitive textile making
8. Early textile of Peru
9. Bark cloth

Chapter 8
1. Sumerian goldsmith's work
    (a) dagger discovered at Ur (Third Early Dynastic Period)
    (b) gold tumbler and bowl
2. Aztec goldsmith's work
3. Ziggurat
4. Mexican step pyramid

Foreword

This book is primarily for children in the 11-plus group; but some of them may wish to discuss its contents with their parents or teachers. For further information the parent or teacher will find the contribution of Jacquetta Hawkes in *Prehistory and the Beginnings of Civilization** an invaluable source of reliable material vividly presented.

Lancelot Hogben

**Prehistory and the Beginnings of Civilization* by Jacquetta Hawkes and Sir Leonard Woolley (Allen and Unwin)

1 The age of our earth and our ancestors

This book is about history, but not about the sort most of us learn at school. When folk talk of history, they usually mean what we can learn about the past from books, manuscripts, scrolls and inscriptions on stone monuments or on tablets of sun-baked clay. In short, they mean what we know about how human beings have behaved since the first of them began to write. Another sort of history, called archaeology, probes into the past by unearthing remains of buildings, weapons, sculpture, tools and the like.

Even when there is a written record by those who made or used such objects, their relics may tell us more than we can learn from it. Faulty memory, credulity, superstition, lack of interest and deliberate untruthfulness can each make the written record very unreliable. Some civilizations have left us a written record in languages we cannot as yet translate or in writing signs we cannot as yet decipher. Despite this, we can learn very much from their remains.

In the centuries between 3000 and 1000 B.C., ancient Crete had successively three sorts of writing. Only during the last twenty years have scholars been able to decode one of them. Thanks to this, we now know a little concerning the language Cretans spoke in 1500 B.C. or thereabouts. For almost everything else we know about them, we have to rely on their architecture, drainage system, wall paintings and other remains which have withstood the test of time.

The other two Cretan batteries of writing signs remain a riddle, and the same is true of the writings of the Etruscans who flourished in Italy between 1000 and 400 B.C. Scholars have now decoded only a few words of their dead language; but we know at least as much about their own way of life as we can hope to learn from

2 Beginnings and Blunders: Before Science Began

1. Homo diluvii testis

Latin records of the way of life of their Roman neighbours before 300 B.C. Their sculpture and burial sites tell us far more than we are likely to learn from their inscriptions when we can decode them.

Writing of a sort began first in Iraq and Egypt. This happened about 5000 years ago. We now know that there have been on earth people like human beings, as we now know them, for at least six and possibly sixty times as many years. For all we can ever hope to know about them, we have to rely on relics of the sort from which we are able to gain knowledge about people whose written records of their past are mere fables or still undeciphered.

Before there were beings who could leave to us in writing a record of their doings, they had to make for themselves a settled life with leisure to write and to read. They could not have learned to do so while they still had to wander far afield and camp where there was wild game to hunt, roots to dig up and fruit to pluck in season. A more settled way of life was possible only when they had learned to keep flocks and herds or to scatter grain and to harvest it near their homes. Only then could they first have permanent dwellings.

Only when there was leisure for building permanent dwellings did people have the incentive to teach themselves skills which make the art of measurement necessary. Only when they had a settled way of life had they leisure to count reliably how many days elapse between

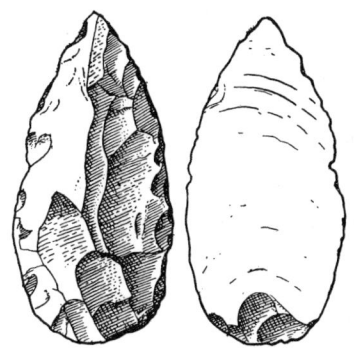

2. Palaeolithic implements
   (a) Old World—
      (i) flint implement of Mousterian type

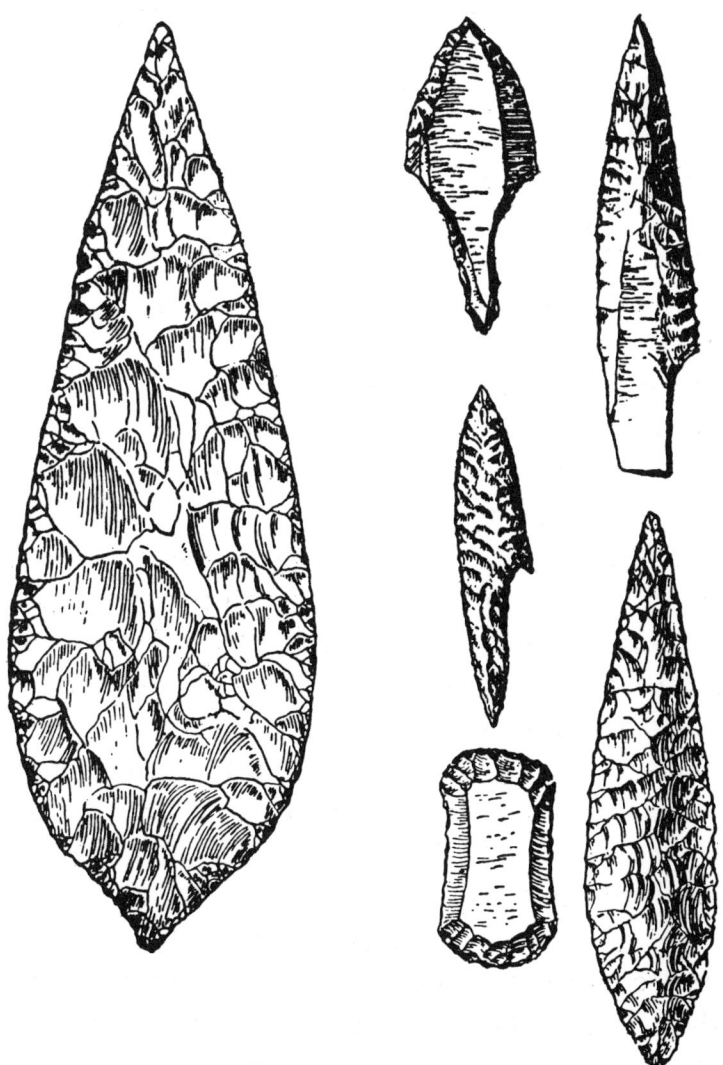

(ii) Solutrean flint implements

4 Beginnings and Blunders: Before Science Began

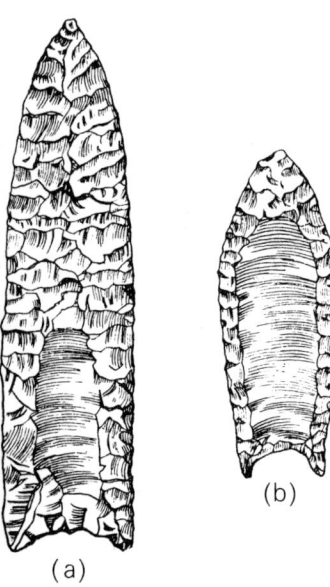

(b) New World—
(a) Clovis point
(b) Folsom point

recurring changes of the night sky, to locate the rising or setting position of the heavenly bodies and to measure the length of the sun's noon shadow. Only when they could record such observations can we say that science had begun.

To be sure, we do not all use the word science in exactly the same sense. Indeed, few people mean the same thing by science when they speak of medical science and Christian Science. Even professional scientific workers can disagree about what they include when they talk about it. None the less, they mean at least this: *science is the written record of reliable knowledge about nature, including human nature.*

Part of the knowledge we have, or would like to have, about human nature, concerns how men and women made it possible for science in this sense to begin. How they did so will be part of the story of this book; but there is more to it than that. To understand how human beings created the settled life which brought science to birth, we need also to understand how they came to be a world-wide species.

Human beings now live on all parts of the earth. They have ranged so far afield only because they are able to protect themselves from the cold. Before they could do so, they had to clothe themselves. Even before they had clothes of any sort, they had learned to make fire. If we want to know how our ancestors first made fire, we shall have to dig into the past to a time far more remote than that of the first appearance of human beings like any we meet today.

A quarter of a million and more years ago, there existed on earth now-extinct creatures more like apes than human beings as we know them in the flesh, but far more human than the chimpanzee, the gorilla or the orang-utan. They were our ancestors and the first creatures to make fire. Less than half a century has gone by since discovery of crude stone tools and ashes with their skeletons gave us our first glimpse of what these near-men looked like. Barely a quarter of a century has passed since atomic chemistry first provided tests to decide how long ago they died out or first appeared on earth.

Almost within the lifetime of people now living, scientific discovery has thus painted for us a new picture of our past. A hundred years ago, most educated people where the Protestant faith was prevalent believed that life on our planet began about five or six thousand years earlier. The Almighty had then made the first man and, from one of his ribs, a woman to be his wife. An English ecclesiastic who devoutly believed in the literal truth of every word in the Bible had dated the event as at 9 a.m. on September 12 in 3298 B.C.

Where people held such beliefs, many still clung to the conviction that the Bible allegory of the Great Flood is also a historical truth. According to this parable, a deluge of rain, which submerged all dry land some five thousand years ago, destroyed all creatures then living except the mythical family of Noah and a representative troupe of animals which accompanied them on a suitably constructed ship.

Before the American Revolution, no more than two centuries ago, people in Protestant countries with pretensions to piety therefore saw in the hard parts of dead animals embedded in rock, slate, coal and limestone, testimony to a devastation due to Divine displeasure with the depravity of Noah's contemporaries. In the reign of the grandfather of George III, the identification of such remains was rarely, if ever, correct.

6 Beginnings and Blunders: Before Science Began

In 1726, a Swiss naturalist displayed a picture of a giant salamander discovered in material we now know to have solidified to rock many million years ago. He called it *Homo diluvii testis* (Man: witness to the Deluge) and placed beneath the sketch a solemn warning. In English one may render it thus:

Of Relics sad of Bone, Frame of poor Man of Sin
Mellow the Heart and Mind of later sinful Kin.

Towards the end of the same century, canal construction went on apace in Britain to meet the challenge of steam-powered production. Surveyors and others then began to study the earth's crust systematically. Such studies piled up more and more likely reasons for regarding the formation of its successive layers as the result of two processes. One is erosion of soil with

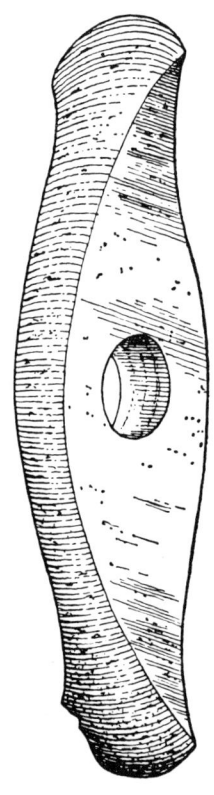
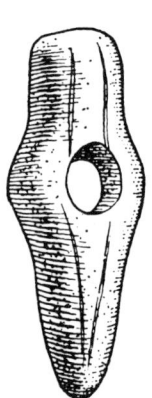
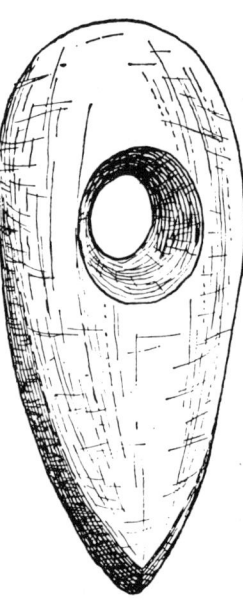

3. Neolithic implements (a) double-edged axes and hammer axe of polished stone

continuous but very slow wear and tear by weather. This process results in submergence of former land masses and elevation of the sea bed to produce new ones; but elevation of the sea bed to form a new layer of dry earth can happen in another way. The sea teems with minute creatures with calcareous (chalky) or siliceous (glassy) shells. At death, the shells sink to the sea bottom, forming a crust which very slowly but surely raises its level.

Except when floods produce large-scale erosion of soil, neither process produces a visible effect in a life-

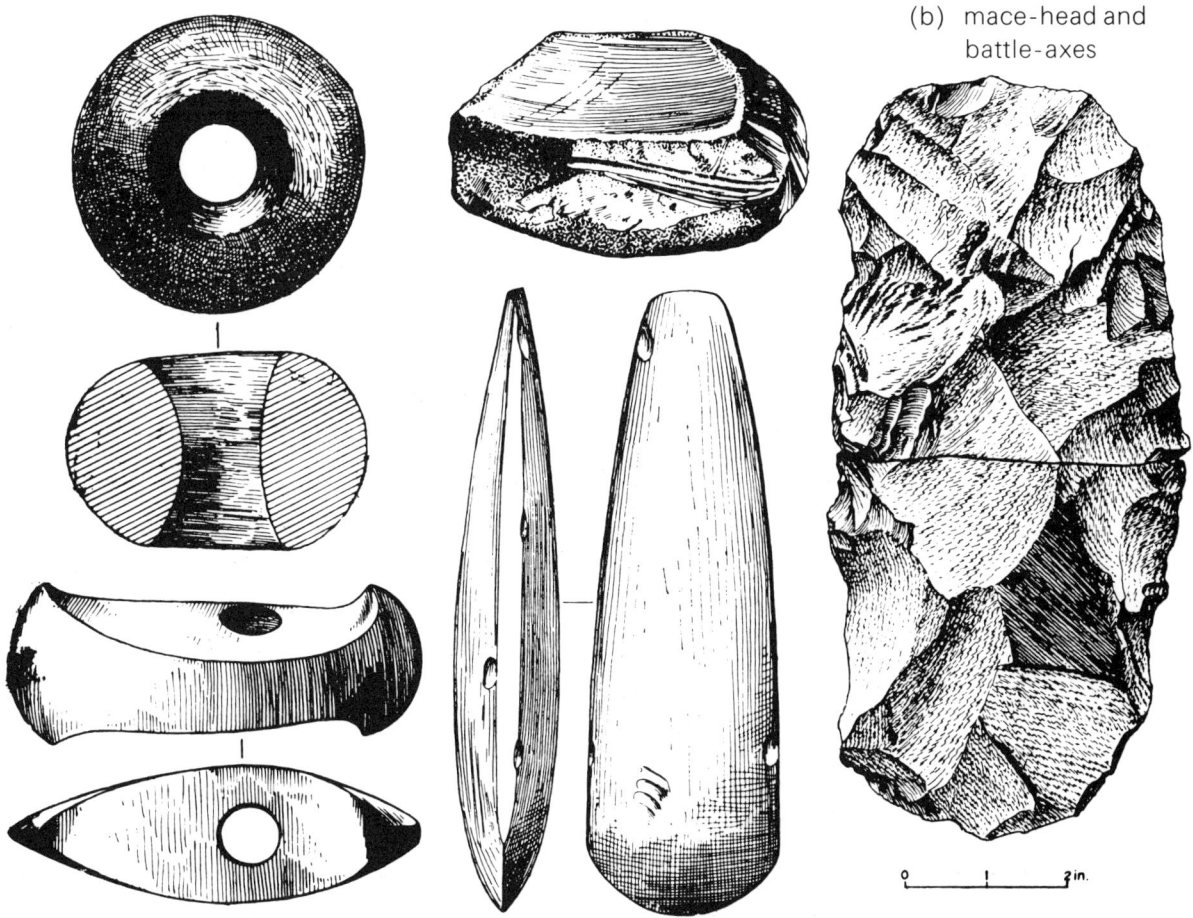

(b) mace-head and battle-axes

time. To have produced the successive layers of the earth's crust, they must have gone on through periods of time vastly greater than the six days of Creation as recorded in the Hebrew Scriptures or even the estimated five or six thousand subsequent years of Protestant theology. When this was clear, fossil remains took on a new meaning. They disclosed a very slow process of replacement of species of animals and plants by others, many of them more like ones alive today.

Many readers of this one will already know that Charles Darwin published in 1860 a book called the *Origin of Species*. In it, he marshalled a considerable array of facts which made it very difficult to stick to a literal belief in the story of Creation, the Great Flood and the origin of our own species as told in the Book of Genesis. Darwin and his followers put forward the view that man is no exception to the rule that the vast variety of living beings, animals and plants alike, is the outcome of a gradual process extending over millions of years, during which the offspring of similar parents have become more and more different.

We express this view by saying that Man and other animals have a common ancestry. In this sense, we may therefore speak of different groups of animals as more or less closely related to Man. When Darwin's doctrine was a novelty, the great apes were the nearest known relatives of Man. The only strong evidence for the belief that there have been animals on earth for millions of years was then the apparent formation of different layers of the earth's crust in the ways mentioned above.

Since his time, study of radioactivity has made it possible to date with confidence the age of rocks and therefore to tell us how long ago lived creatures which have left behind in them hard remains of bone, shell or

4. Australian saw-knife

wood. Meanwhile, many such relics unknown to Darwin's generation have come to light. While we have thus learned more and more about Man's antiquity, and about his nearest relatives, we have also gained more and more knowledge of tools and weapons used by human beings like ourselves before the dawn of history and by their ancestors before human beings like those of today appeared on the scene.

They made their first tools and weapons by chipping stones, especially flint or quartz. Later, they learned to use bone, ivory and the antlers of stags. Until the beginning of our own century, the study of the changing pattern of such instruments gave us what clues we had to the increasing command over its environment by our own species or by its tool-making ancestors. Pioneer workers in the field thus occupied themselves largely with the methods of chipping, boring and grinding employed to fashion them.

5. Australian using hand-axe

Though fascinating to the expert, the techniques employed for their manufacture are of less interest to most of us than are other discoveries which help us to understand the way of life of the users, and what they used them for. None the less, there is one question which every curious reader will ask or will have asked: how do we know that chipped or seemingly polished pieces of rock are *artefacts*, i.e. fashioned by men or near-men?

To answer it, we need to know that such rocks or stones are of a material which flakes to expose a sharp cutting edge when hit hard or subjected to strong local pressure. Frost, ice, heat and the crushing power of waves can do this. However, such natural agencies cannot produce patterns like those which the human hand can produce.

There are indeed living people in Australia who still practise this art, and there were far more until the native population of Tasmania died out in the nineteenth century. Among such, stone arrow and spear heads, axe and knife blades secured to wooden shafts or handles by twisted fibres of tree bark or strips of animal hides, have been in use during recorded time. This has given experts who excavate the buried relics of our past good grounds for classifying stone implements by the uses to which our ancestors have put them.

Details of their manufacture and discussion of borderline cases whose manufacture by hand is in doubt may well be tiresome to us unless we have opportunities for digging where human or near-human artefacts abound. What most of us want most to know about our ancestors who lived long before writing began, and what this story of Beginnings and Blunders is about, is what sort of life they lived. The following questions are some of those to which we shall seek an answer.

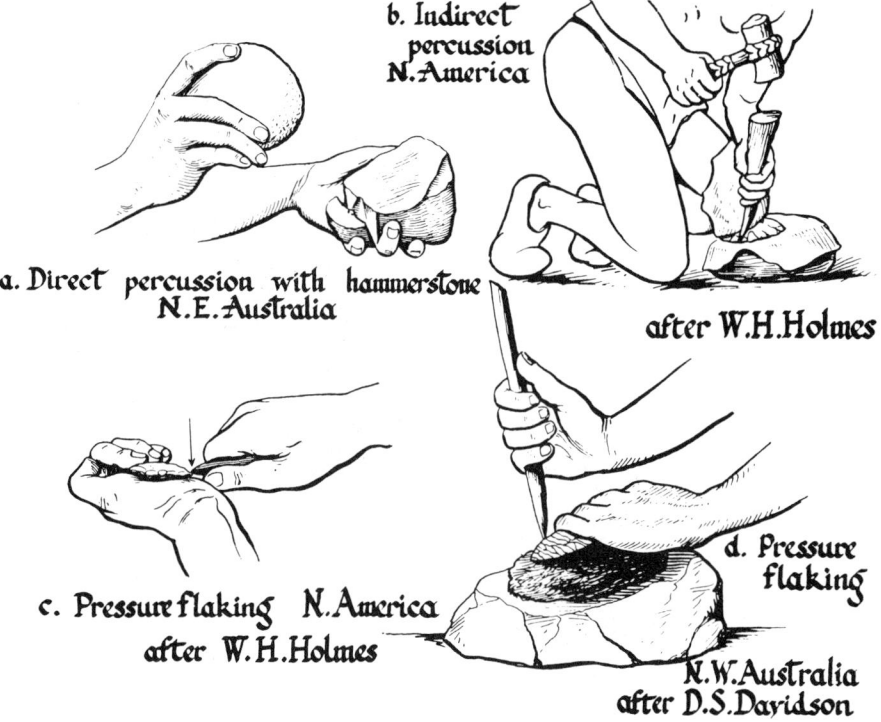

6. Methods of making tools

a. Direct percussion with hammerstone N.E. Australia
b. Indirect percussion N. America after W.H. Holmes
c. Pressure flaking N. America after W.H. Holmes
d. Pressure flaking N.W. Australia after D.S. Davidson

By what skills did our remote ancestors learn to penetrate into regions too cold either for survival of their naked bodies or to provide them with sustenance while still wholly dependent on hunting, fishing and gathering edible roots, fruits or seeds? When and how did they first learn to fell trees and to make boats? How did they come to keep goats, sheep and cattle alive. How were they able to breed them in captivity instead of pursuing and killing them? How did they come to sow crops? How did they learn to make pottery and to use metals? How and when did they acquire the arts of making baskets and of weaving fabrics? How did trade between the first settled communities come into being? What made possible the first sort of city life?

In so far as we can outline a time chart setting out when and where this or that happened we have to rely

12 Beginnings and Blunders: Before Science Began

7. Cave painting of bow and arrow hunt

on bones, stone tools, scraps of pottery and remains of dwellings buried, till excavated, under soil deposited by erosion or by sand blown during sandstorms. Since our aim will be to study only the human story before the written record begins, we may divide it, as expert diggers do divide it, into two periods.

Earliest, and lasting 200,000 years or more, is the Old Stone or *Palaeolithic* Age before settled village life began. During this period our ancestors used tools and weapons made for the most part of flaked stone or bone. With the coming of farming and village dwellings they had begun to use smoothly polished stone for axe-heads, sickles and so forth. We speak of this period as the New Stone or *Neolithic* Age. The earliest known Neolithic communities date from about 8000 B.C.; but when we put dates to events in this way, we should always bear in mind three things.

One is that a discovery made tomorrow may elsewhere unearth a similar level of development of Man's body, tools and social habits at a much earlier date than the earliest date available information now assigns to it. That discoveries made in parts of the world little studied so far may change our ideas about the sequence of events is not unlikely. For every square mile excavated so far, there are tens of thousands of square miles where there has been no digging. We know as yet vastly more about what is below the soil in Egypt, the Middle East and Europe than we do of what excavations of vast tracts of territory in tropical Africa and tropical Asia may one day uncover.

Another thing we should always bear in mind is that the word *Age* can be misleading when we are talking about different regions. When one speaks of the New Stone Age beginning, or the Old Stone Age ending, at such-and-such a time, such statements mean nothing unless applied to a particular territory. When Europeans came to Tasmania, the aborigines were men of the Palaeolithic way of life. The Neolithic way of life which was well established in Iraq by 7500 B.C. did not reach Britain much before 3000 B.C.

A third thing to remember when we confine our study of the past to relics we recover from sifting soil or sand where human beings have once camped or made more permanent settlements is that some things resist decay and weathering far more successfully than others. Stone tools are more durable than bone ones. Earthenware lasts longer than basket work or leather. At a particular site we may find pottery before we find traces of basket work. If we do so, we should not necessarily conclude that people who camped or settled there found out how to make pottery before they had learned how to make baskets.

In more than one way, a knowledge of biology should keep us alert to this source of error in the attempt to recapture a remote past of our species. When discussing the evolution of a particular group of animals and plants, biologists give most weight to the evidence of fossils, i.e. bones, shells, wood, impressions of leaves, feathers, footprints, etc, but they do not close their eyes to what resemblances between closely related living forms and their present geographical distribution can teach us.

In the same way, the science of animal and plant breeding, the study of where we find particular species living today, and what we know about their habits, can contribute to the knowledge of our past, information which we can never hope to gain from excavation. This is especially true of what archaeologists call the domestication of animals and plants.

As the human story before the written record unfolds, we shall repeatedly have occasion to cite an approximate date to events and relics of our past. So a few words about how radio chemistry has lately come to our aid are fitting at the outset. To understand how, we need to know that the lighter elements have unstable forms which emit rays like the more stable heavier ones, i.e. thorium, uranium and radium.

Though not recognizable by chemical, they are detectable by physical tests which disclose a peculiar property of living matter. That is to say, animal and plant tissues contain during life a remarkably constant proportion of radioactive carbon, called carbon 14 in contradistinction to the more stable and abundant carbon 12 of school chemistry.

Tests for radioactivity have taught us that carbon 14 gradually deteriorates to carbon 12 and have established the rate at which it does so, so the proportion of carbon 14 remaining in the carbon content of fossilized wood, bones, shells and peat at excavation sites enables us to estimate how long ago such relics ceased to be part of a living plant or a living animal.

2 What our ancestors were like

The animal species which human beings resemble most closely are monkeys and apes, more especially apes, which are larger than monkeys and have no tails. Among the ancestors of such animals we therefore expect to find our own. To talk about what they were like we shall need names for them, and to do this we must first be clear about what biologists mean by *species*. The use of the term is not as clear-cut as one might wish. When we speak of animals or plants as members of a particular species, we mean a group of creatures having in common certain noticeable characteristics which they do not share with any other group. We also mean that they breed freely among themselves but not with members of other species.

To say that members of the same animal species breed freely means that they mate and produce *fertile* young. That members of one species do not breed freely in this sense with members of another may be due to one or other of three circumstances. As is true of dogs and cats in the same household, they may have no disposition to mate when within reach. As is true of the horse and the donkey, some may mate and produce cross-breeds (called *mules*), which are sterile, i.e. unable themselves to have young. As is true of different species of tapir, one found only in Malay and the other in South America, they may have no opportunity to mate.

If we know that members of two such distinguishable groups of similar creatures could produce young when brought together, as in a zoo, it is not unusual, and certainly better, to call them geographical varieties of the same species. Unfortunately, we often lack such knowledge. The distinction is not the less important, especially when we discuss the so-called domestication of animals. When we ask what was the ancestor, or

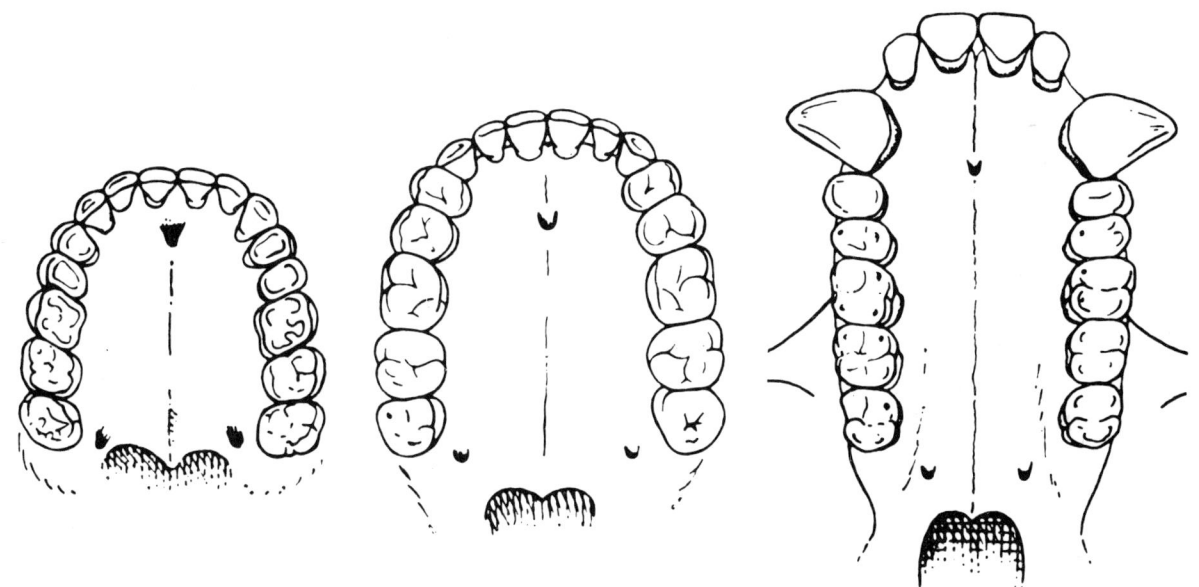

1. Palate of *Australopithecus* (*middle*) compared with man (*left*) and gorilla (*right*)

were the ancestors, of our domesticated breeds, it is not good enough to label one or another by the name for a species unless the name stands for a group of animals which either do not mate with members of other groups when opportunity arises or, if they do so, produce only sterile crossbreeds.

Living human beings from different localities have more or less distinctive characteristics, but they can interbreed freely and we therefore use the word in this sense, when we speak of them as members of one species. Biologists give all animal and plant species two names: a surname and a given name, both derived from Latin or Greek. As in a telephone directory the surname comes first. The full name of our own species is *Homo sapiens*. The surname (also called *generic* name) *Homo* is Latin for Man. Several species which share characteristics which they do not share with others make up a *genus* with the same surname. Thus *Equus caballus* the horse, *Equus asinus* the donkey, and *Equus zebra* the zebra, are all members of the genus *Equus*.

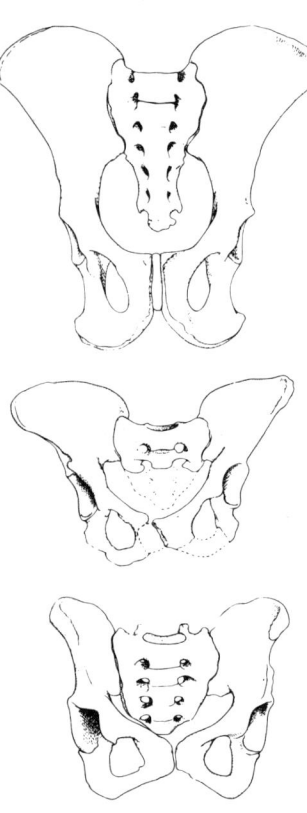

2. Pelvis of *Australopithecus* (*middle*) compared with man (*below*) and gorilla (*above*)

Though there is no hard and fast rule about which species we should group in the same genus, no zoologist would place our own species in the same genus as any ape. The differences between Man and any living ape are far greater than the differences between species of the latter. Aside from the fact that the size (i.e. volume) of the brain of the largest living ape (about 500 cc) is much less than that of our own (average 1350 cc), many other physical characteristics are peculiar to Man.

Some involve the palate and teeth. Even more striking are others which involve his mode of progression, and his abandonment of tree dwelling. The arms, relative to both trunk and legs, are shorter than those of any ape. The bones of the legs are not bent and the structure of the foot is like that of neither monkey nor ape. In particular, the great toe is firmly united with the sole. It is thus incapable of grasping the branch of a tree, as can the free great toe of monkeys and apes. Unlike theirs it is also larger than the other four.

There are four living types of ape. Two species, the chimpanzee and the gorilla live only in tropical Africa. The orang-utan and the gibbons live only in tropical Asia. Excavation in tropical and subtropical Africa during the past fifty years has brought to light skeletons of creatures of comparable stature and with brains appreciably bigger. Till more of their remains have turned up all experts will not agree about the last statement. At present, most of them agree that they were in several ways more like human beings than any living apes. Their hips were relatively wider and shallower than those of apes. Their palate was relatively broader. The canine teeth were less prominent. The heel was more like that of Man.

Anatomists are not unanimous about whether some of

these ape-men, should have the same surname *Homo* as ourselves. They call them *Australopithecines*, meaning southern apes though no apes live so far south as Australia. An up-to-date estimate of the period to which their remains belong is between two million and half a million years ago. Somewhere about the time when they became extinct, far more human creatures left their remains in Java and in China. Though somewhat smaller than ours, their brains were much larger than those of a living ape. Unlike ourselves, they had prominent ridges above the eye socket. From their limb bones, we know that they walked erect.

Their discoverers named the Javanese specimens *Pithecanthropus erectus*, which means upright ape-man, and put the Chinese specimens in another genus *Sinanthropus* (China-man). More recent workers have named them *Homo erectus*. This raises the question: where should we draw the line between species to which we do and do not agree to give the surname *Homo*? As matters stand, we can say that the forms called *Homo erectus* certainly share one unique characteristic of living people.

3. Skulls of chimpanzee (A), *Homo erectus* (B), and *Homo sapiens* (C)

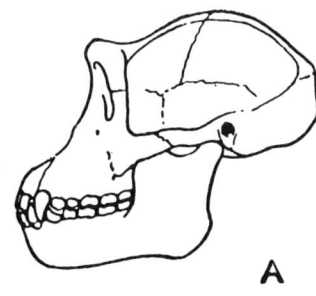
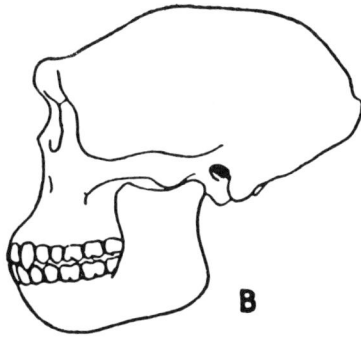
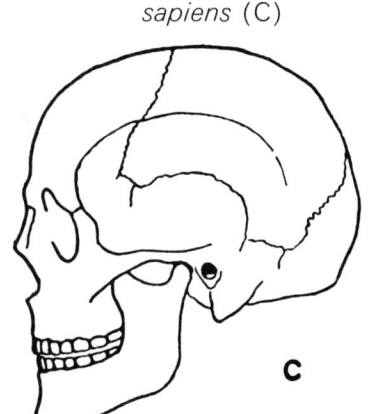

The behaviour of the human species is unique in two ways. As Benjamin Franklin pointed out, Man is a tool-making animal. Man is also the only animal that can communicate by speech. Parrots, budgerigars and mina birds can make noises like human words or short phrases; but they are not able to make noises which tell their offspring where to look for the next meal. Man alone can use words to hand on the know-how of his tool kit to his successors. Because of this, each generation starts with the benefit of experience gained by preceding ones.

Their skeletons alone can tell us little or nothing about whether our ancestors were talkative in this sense. Whether *Homo erectus* was we cannot say. We do, however, know that he was a tool-maker; and to that extent alone he was more like ourselves than any other living creature. To be sure, the tools he made were very crude; and there is room for doubt about the precise use he made of them. None the less, we can be certain that they were man-made.

The Chinese skulls and other fragments of the skeleton, including teeth from as many as forty individuals, were discovered in a cave side by side with chippings of quartz. Such chippings could not have come there unless brought from outside. No animals other than men—or near-men—would have done so. All the indications are therefore that *Homo erectus* used the shelter of the cave as a family factory for flaked tools. We cannot with certainty trace the first stages of this break with our animal past; and there is no present prospect of being able to do so.

However, we do have a few important clues which suggest that our *Australopithecine* ancestors had some sort of crude tools. From remains found with them, it

What our Ancestors were like   21

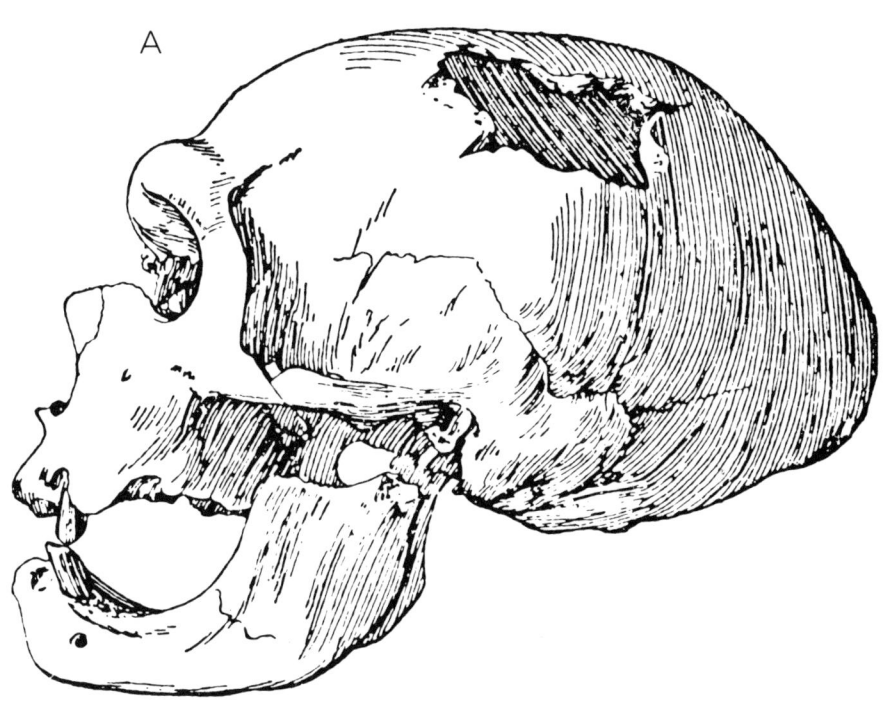

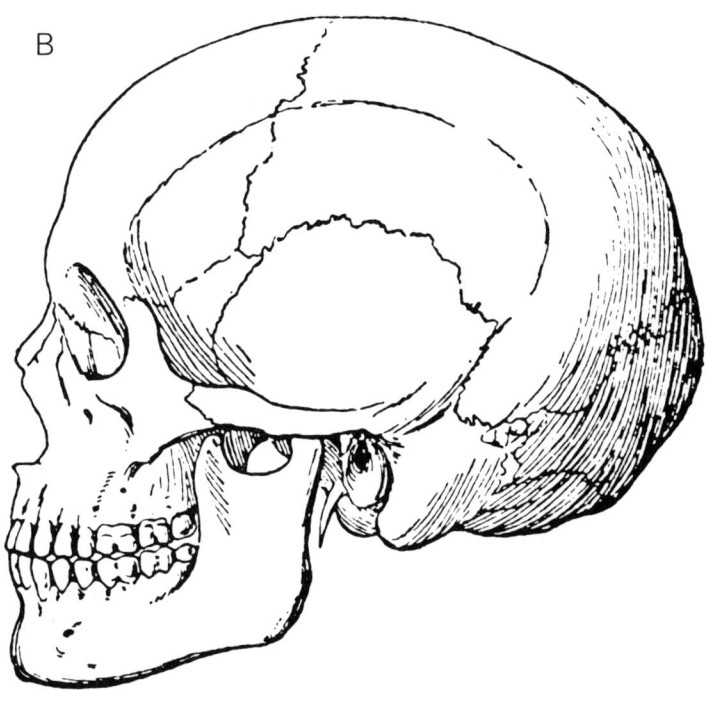

4. The Neanderthal skull (A) compared with a modern human skull (B)

seems that they were not almost exclusively vegetarian like the apes of today. If hunters and meat-eaters, they had to catch their prey and dismember it for digestion. It is unlikely that they could run fast enough to catch their prey bare-handed, and if they did so their teeth, like those of *Homo erectus* and of ourselves, would be unequal to the job of tearing away the inedible pelt to expose the succulent flesh of prey such as deer or other species they hunted.

Certainly *Homo erectus* was a meat-eater. For we find with him in the family flint factory charred and broken bones of wild animals. Pebbles found with the remains of some of the *Australopithecine* species may have been used as missiles in the chase of game. Such remote ancestors may also have used sticks as weapons. If so, they have left no trace. Whether or not they did so, flaked stones with cutting edges would be more effective missiles than smooth pebbles; and they would have another advantage when the hunting family group trapped or maimed their victim for the kill. The cutting edge could be the means of scraping the hide from the carcass, of cutting tendons binding meat to bone and of splitting the bones to extract their marrow.

Homo erectus comes on the scene well over a quarter of a million years ago. During a period of a hundred thousand years after we lose trace of him, very scant bony remains of others among our ancestors have turned up so far. Digging and exploration of caves has, however, exposed a succession of more and more finely worked, and hence more and more clearly man-made, products. With one stage in this progressive elaboration, the so-called *Mousterian* culture, we find remains of a species ranging from Africa to Asia and to Western Europe.

Their discoverers speak of European representatives

of this species as *Homo neanderthalensis* (Neanderthal Man). They flourished between 100 and 30 thousand years ago. The names Neanderthal (from a Rhine valley) and Mousterian (from Le Moustier in France), refer to the localities where their remains have turned up. As found in Europe, Neanderthal Man seems to have been the dead end of a side branch of our ancestral tree. Since his thigh bones were somewhat bent, he probably had a slouching gait. His forehead and the front of the lower jaw sloped backwards like that of *Homo erectus*. Like the latter, and unlike ourselves, he had very prominent bony ridges above the eye socket.

From his limb bones, it seems as if he walked less upright than *Homo erectus*. Partly for this reason and partly for others, we cannot place him as on the direct ancestral line of our own pedigree. Whether he did or did not interbreed with contemporaries more like ourselves, we cannot say. A South African fossil (*Homo rhodesiensis*) with a similar skull but with more upright bearing is probably closer to our own ancestral tree trunk.

Two remains of wooden shafts from widely separated periods of time show that he or his near contemporaries used the spear. Likely enough also part of his tool kit, are a digging stick and a club found in Rhodesia, and dated as about 60 thousand years ago. How we date such relics, we already (pp 14–15) hold the clue. We know that Neanderthal Man habitually used fire. There are indications that he buried his dead beneath the hearth. If not devout, he was at least a mortician.

The later Neanderthal men had spread northward from Germany where the climate was becoming more and more severe during the period to which their remains belong. It is therefore unlikely that they could have

5. Rhodesian man

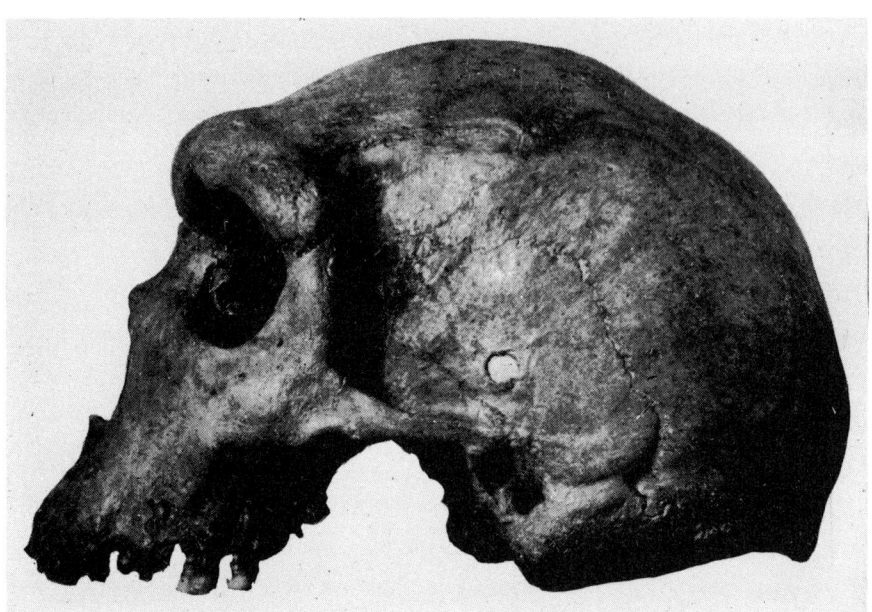

survived without using the hides of animals to keep out the cold. When not engaged in seeking prey, they sought shelter under rocks and in caves. Bones charred and split for extraction of marrow show that they cooked their meat.

They must have been skilful hunters. Since their prey included such large animals as the woolly rhinoceros and the extinct elephant known as the mammoth, we can be sure that they had learned an art still practised by African hunters. They must have known how to dig deep pits concealed by branches of trees and to lure their prey to them. Since we find with their remains many long bones with very few, if any, vertebrae or ribs, we can also conclude that they cut up the carcass without attempting to drag it to their lairs.

Shortly before and after the disappearance of remains associated with Mousterian flint instruments, and presumably after extinction of their makers between thirty and forty thousand years ago, successive waves of more inventive hunters from Asia penetrated into Europe. They made a greater variety of tools of finer workmanship—so much so that we can have no doubt about their uses. Some were spearheads, some were knives, some arrowheads. Their makers could fashion pins and awls of bone and harpoons of stag antlers.

Such hunters who came into Europe immediately after we lose track of the Neanderthalers left on the walls of caves paintings and engravings remarkably like those which the Bushmen of South West Africa have continued till today to make on rock faces. From the same sites as their tools, the few skulls unearthed so far are indisputably like those of people now living. Whether men and women like ourselves were the makers of earlier and more crude tools, we do not as yet know certainly.

The cave pictures made by our ancestors during a period of 20 to 40 thousand years ago are fascinating as art, sometimes of great beauty and accuracy. Some, the more attractive, depict animals. More crudely, others display incidents, for instance: men fighting with bows and arrows; pit-digging to trap prey; a woman (p 29) using a grass rope ladder to collect honey from a hive on a rock face. Thus the cave pictures confirm much we might otherwise merely guess to be the uses of the stone and bone implements of their artists.

Ten thousand years or so before a more settled way of life gave more scope for inventive skill, other varieties of our own species seem to have replaced the Picture Makers of what archaeologists call the Old Stone Age in Europe. During all the period between the coming of our own species and the beginning of village life, human beings were nomads. They wandered from place to place to find food in season. Since the Picture Makers certainly, and very probably the Neanderthalers also, could

6. Early cave paintings

dig pits to trap big game, it is highly likely that they could dig up edible roots.

Where they spread to the sea shore, they could also find shellfish at suitable sites. Fish depicted on some cave walls, and the fact that the later Picture Makers made harpoons, indicate that fishing contributed to their menu. This must have been an immense advance in the struggle for survival. Except when frozen in the far north, rivers supply a source of food at all times of the year. We shall later see how this pointed one way to settlement in more permanent dwelling places than the camp sites of their predecessors.

The harpoon is a spear with a barbed piercing blade and a cord by which the hunter can regain it. That our ancestors seem to have used it before they made fish hooks from bone or nets from rope will come as a surprise only to town-bred folk who have never seen poachers land salmon with a gaff. The harpoon is an elaboration of Man's earliest weapon for hunting big game. As a means for catching large fish an ordinary spear could be efficient only when the piercing edge had barbs to ensure that the prey could not wriggle free; and only a cord attached to it could then make it impossible for the fish to swim away with the weapon in a dash for freedom.

3 Food, fires and fur coats

We have seen that our early ape-like ancestors, the *Australopithecines*, were not strict vegetarians. They had acquired a taste for meat and that has a bearing on where human beings now live. Being vegetarians, the apes of today live within the tropical belt, where fruits of one sort or another, such as coconuts, paw-paws and bananas, are available at all times of the year. Enjoying a mixed diet, our own species has established itself in every part of the world, from the north polar region to Antarctica.

If our remote ancestors had shunned meat, they would have had no incentive to fashion weapons for hunting, and they would have had to remain where a vegetarian diet was always available. The search for wild game was an invitation to migrate far beyond their tropical homeland wherever there was meat to offset seasonal shortages

1. Cave paintings of animals

28 Beginnings and Blunders: Before Science Began

2. Cave paintings of women's attire

of fruit, nuts and berries. To ensure their survival and that of their descendants in cooler regions, they had to improve their hunting methods. Only greater prowess as hunters could permit them to multiply without exhausting food supplies.

During more than a quarter of a million years since *Homo erectus* appeared on the scene, life at any considerable distance outside the tropics forced men or near-men to face another challenge. Geological study has shown that the earth's climate during this period underwent vast changes. Four times the polar ice cap has extended far beyond its present limits, engulfing in ice parts of Europe, Asia and North America which now enjoy a mild climate and again did so each time the ice receded. During the last of the ice ages, reindeer were abundant in northern Spain, and hunted by folk of our own species.

Such drastic climatic upheavals repeatedly faced our ancestors with the choice of either retreating from

Food, Fires and Fur Coats 29

territories they had already peopled or equipping themselves to remain. To retreat into territory peopled to the limit imposed by depleted food supply might have led to extinction. To stay on was practicable only if they developed new skills to cope with changing resources of animal and vegetable food; but new expertise as hunters and gatherers of edible plants could not suffice to guarantee survival. Our ancestors had also to protect themselves from extremes of temperature by making their own climate.

The story of how they did this begins with the use of fire. From the dawn of history people have speculated about the origin of this peculiarly human accomplishment. An ancient Greek myth relates that a hero named Prometheus brought fire down to earth-born beings from the gods of Olympus. In the early part of our own century, before the remains of *Homo erectus* and of his cooked meals came to light near Peking, three theories, which now seem almost as fanciful though not as poetic, were widely current.

One suggests that our ancestors first obtained fire by making bow-string-driven hand drills of hard wood to generate enough heat to set dry moss or wood alight. A

3. Cave paintings of human behaviour

second claims that they first learned to use fire by scooping up red-hot lava from erupting volcanoes. According to a third, they took advantage of embers left by forest fires started by lightning.

Of such guesses, the first is the least plausible. To be sure, some people, cut off from civilization in extremely dry climates where very hard wood is available, have learned to make bow-string drills. Friction generated by drilling long enough can set dry leaves or moss glowing to kindle a fire for cooking; but those who go to the trouble of making and using such a tool do so with the firm assurance that a fire will be useful to them once they have made it. It is incredible that near-men obtained fire by such a laborious process before they had any notion of its usefulness. Further, there is no reason to think that they knew how to make bows or bow-string drills.

The forest fire theory makes sense only if we assume that the near-men who first kindled a flame for use were very different from all other wild animals. Almost certainly, they would flee in terror from a forest conflagration; and, even if bolder than is likely, they would rarely get the chance to see one. Where such fires now occur, they are usually due to barbecues or careless cigarette smokers. Except perhaps in the eucalyptus forests of Australia and some pine forests of America, forest fires caused by lightning are rare occurrences; and we know that our ancestors were making fires long before they peopled either continent.

Much the same objections apply to the red-hot lava story. A volcano belching forth flame, smoke and lava is not a sight to inspire near-men to meditate on the possibility of making better use of fire. Besides, there is little or no evidence that the first of our ancestors to learn the trick of making a fire for warmth or cooking lived in the vicinity of active volcanoes for any length of time, e.g. Peking.

The charred remains of meals found with those of *Homo erectus* in the caves near Peking prove conclusively that creatures more ape-like than ourselves had already learned to make fire long before *Homo sapiens* arrived on earth. In seeking for a reasonable explanation of this discovery, it is therefore unwise to exaggerate the inventiveness of the first of our ancestors to do so. When we speak of inventiveness today, we mean finding better ways of doing something we have already learned to do somehow. We do not mean producing new tools or new machines for which we have not yet thought of a use. So it is reasonable to assume that near-men discovered the usefulness of fire before they deliberately set out to make it; and that they first became acquainted with it in circumstances unlikely to spread panic.

Our master clue to the mystery is that the habit of chipping flints and of making fire go hand in hand nearly as far back as we can trace their ancestry. The impact of flint on flint or quartz on quartz may produce sparks, but such sparks are rarely, if ever, sufficient to ignite material on which they fall. However, it is not uncommon for flint to have bits of harder stone embedded in or near it: for instance, iron pyrites whose name comes from Greek meaning *conversant with fire*. Its impact on flint could produce sparks sufficient to set dry leaves or moss aglow. Gentle blowing or fanning with the hand can then kindle the glow to a blaze, not so hot as to terrify yet agreeably warm, in the chill air at sunset.

Since we know that near-men must have spent much of their time knocking flints together to shape them into tools, it seems likely that the flint factory, such as that

32 Beginnings and Blunders: Before Science Began

4a & 4b Animal totem star clusters

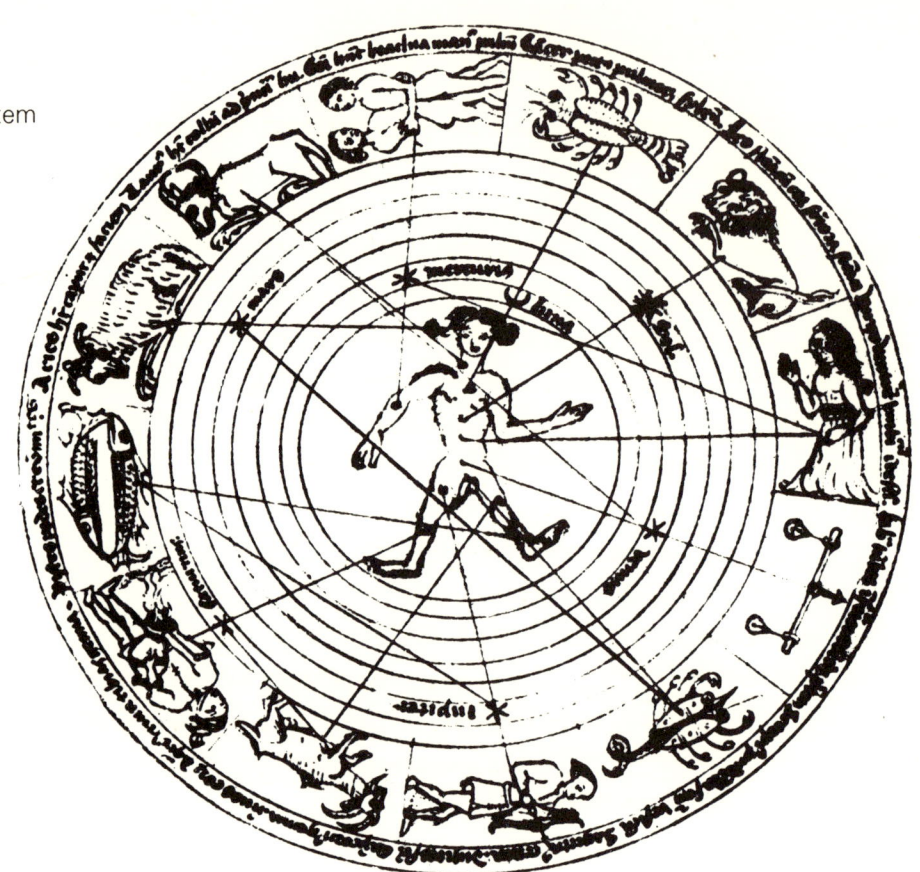

4a

Food, Fires and Fur Coats 33

of Peking Man, was the site of the earliest recognition that fire is useful. Only after a lucky blunder in the course of chipping new tools would near-men be likely to learn how to light small fires deliberately. If so, the method first used was not very different from the one practised in the time of George Washington or Nelson. Before matches came into use a hundred and fifty years ago, the great-grandparents of people still living made a flame by letting sparks from the impact of flint on steel fall on old rag fibres called tinder. Cigarette smokers now use sparks produced in the same way to ignite the petrol or gas in their lighters.

Without fire it is doubtful whether *Homo erectus* could have survived the cold winter nights at the latitude of Peking. *Homo erectus* did so; and we have seen in Chapter 2 that some of his successors, including Neanderthal men and early members of our own species, lived in parts of Europe far farther north during periods when the climate there was much colder than it is now. The use of fire helped them to do so; but it does not fully explain their survival. Tired hunters were doubtless glad to lie down near a warm blaze when darkness fell; but they would have to spend most of the daylight hours

4b

5a

5a, 5b & 5c Animal disguises

on the move in search of their next meal. As they wandered about through biting winds, driving rain, or falling snow, they had no means of carrying a cheerful fire with them. Having become less and less hairy, they had to equip themselves with a borrowed skin.

Had they waited to experience the rigours of northern winters before discovering some way to keep warm out of doors, they would have perished long before they invented one. So it seems likely that they took to wearing apparel before they migrated far northward. If so, it is less likely they first did so to keep themselves warm than that their main concern was with other uses of clothes. A present-day bishop does not wear a mitre to protect his head from the weather, nor does a student wear a gown to keep his shoulders warm. Both sorts of attire proclaim their wearers' place in society. What little we know about clothing in the Old Stone Age suggests that a comparable intention prompted its earliest use.

We have no relics of clothing from such far-off times, nor do we know when our ancestors first started to clothe themselves. All we have to go on are the few cave pictures that show people clad in animal skins. These

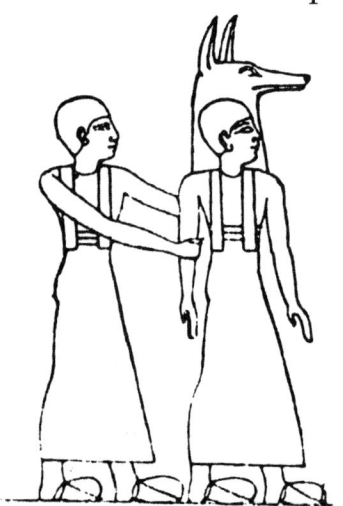

5b

5c

suggest that the Picture-Making people wore them with little concern either for convenience or for the weather. Indeed, such pictures may show a skin hanging from the wearer's head and masking much of his face in a way which would have made it impossible for him to hunt or even to run very far. There is little sign of any gear to keep the lower part fastened around the wearer's trunk or thighs, as modesty or protection from cold would require. So far as we can judge, the skins therefore serve no purpose other than to identify the wearers with certain animals.

As it happens, groups of living people whose way of life is not very different from that of the later Old Stone Age hunters dress up in much the same way with that end in view. Among such people, living in widely separated parts of the world, we meet with two different

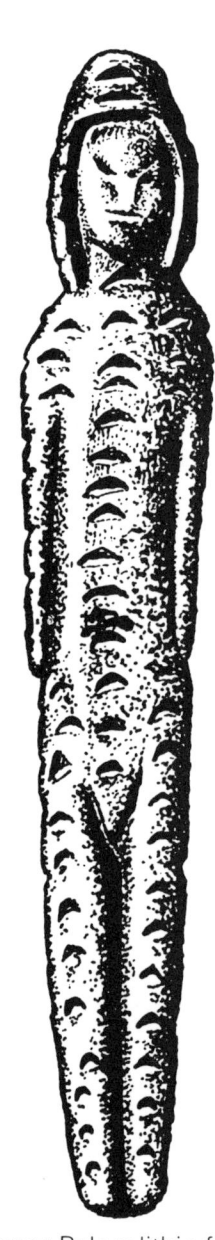

6. Upper Palaeolithic fur clothing—ivory carving

motives for wearing the skin of an animal. One is the superstitious belief that it endows them with the most enviable qualities of its original owner. Thus the man who wears a lion skin hopes to gain the strength of a lion.

A second motive is more difficult to understand. Among people who still live much as the later Palaeolithic Picture Makers lived, we commonly find that neighbouring communities—we may call them clans—hold some species of animal sacred, and each takes its clan name from it. To the Bear clan the bear is sacred, to the Wolf clan the wolf is sacred. We call such an animal the *totem* of the clan, and round the totem cluster some odd beliefs. Except on one day of the year, when a group of stars, which may bear the name of the totem, rises just before sunrise or sets just after sunset at its proper point on the horizon, no member of the clan may slay it or eat it; but when the appointed day comes round, the flesh of the totem is the chief delicacy at a sacrificial feast. At the feast some or all members of the clan wear the skin of the totem animal, and take part in a charade.

There is also another, and probably more compelling, reason for wearing the skin of the totem animal on other occasions. In some communities tribal law dictates that nobody may choose a mate of the opposite sex from among his or her own clan. In others, it is the law that one can mate only with a member of one's own. Either way, the penalty for transgression may be very severe. It is therefore a wise precaution to wear the skin of one's clan totem. When girl meets boy, each then knows whether it is reckless to date.

In short, folk may wear the totem skin only to avoid the penalty of breaking tribal law. When travelling far afield however, they will find another use for it. Sooner

Food, Fires and Fur Coats 37

or later bad weather will come, and they will be glad of it as a protection against cold or rain. Did our ancestors practise such totem rituals with similar rules about choice of mate, and did they first blunder into wearing clothes because of them? Nobody can give a definite yes or no to the question; but we have enough evidence to justify the answer: it may well have happened.

The earliest pictures we have of animals, and of people wearing animal skins, are not rough scrawls found near the mouths of caves, where our ancestors are likely to have bivouacked from time to time. They are works of art, executed with care and without haste. They are mostly found in the deep, uninviting interiors of caves, which could then have been lit only by a burning brand of wood or some crude form of stone-hollowed lamp. Relics of such lamps remain with us. The artist worked therefore in great difficulty.

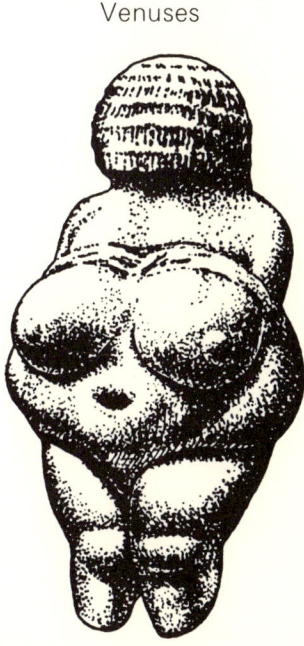

7. Old Stone Age Venuses

Thus everything we know about them tells us both that the artists regarded what animals they depicted as of overwhelming importance in their lives, and that wearing animal skins was an act worthy to place on pictorial record. Both animals and skin-wearing must have played a part in the ceremonial practices of the deep interior of the cave; and the ritual may well have been like those of people now living outside the pale of civilization.

For whatever reason they did so, our ancestors took to wearing clothes well before the end of the last Ice Age in what are now temperate regions. Without them they could never have survived in a cold climate. With flint knives, needles of pierced bone, and sinew or strips of pelt used as thread, people of the late Old Stone Age fashioned for everyday wear in cold weather fur coats and jackets of hide—garments many of us now regard as luxuries. To make clothes by weaving spun animal or vegetable fibre, now commonplace but more comfortable and convenient, called for skills which people could learn, and for materials which could become available for use, only when people had settled down to village life with crops or flocks to tend.

In a cold climate, fire and clothes are as necessary to human survival as food. Alone they are not enough. Shelters of some kind are also essential. If settled in Europe during the Old Stone Age, people like ourselves must therefore have had them; but it is wrong to imagine that they used for permanent shelters the caves in which archaeologists have found some of their remains.

Dry accessible caves are few and far between. Permanent residence in them was possible only if all-the-year-round supplies of nearby food were assured. During the last European Ice Age, Neanderthalers could regularly

Food, Fires and Fur Coats 39

8. Eskimo clothing made of animal pelts sewn together with bone needles

re-stock the larder during winter periods when they were able to trap huge animals like the mammoth and the woolly rhinoceros; and it is therefore possible that some Neanderthalers did become permanent cave-

40 Beginnings and Blunders: Before Science Began

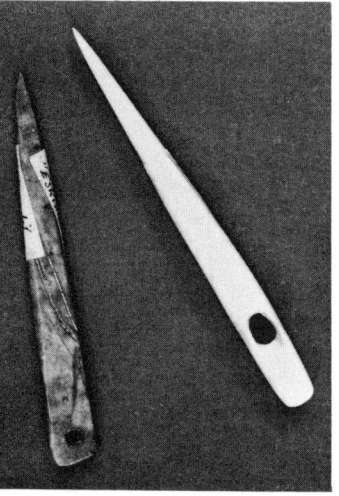

9. Bone needles for making clothing

dwellers. If so, the attempt to stay on in familiar caves after they had almost wiped out the big game of the neighbourhood may have been a cause of their extinction.

Other men and women of the Old Stone Age were compelled to range over very wide areas, hunting much smaller animals, and searching for fruits, nuts and berries in season. For them, permanent cave-dwelling was out of the question. Without doubt, they sometimes sought temporary shelter in the mouths of caves, and it is likely that the same tribal group returned to bivouac at the same site many times in the course of their seasonal travels. They have left untidy remains of their meals trampled into the soil near cave entrances—but never in the deep interior. They must have used shelters of a different sort when out of reach of caves.

Small groups of men may have joined to tear down branches to assemble temporary shelters like those of the natives of Tierra del Fuego, as described in Darwin's *Voyage of the Beagle*. Like Australian aborigines, they could use stone hand axes to cut down bigger branches to make more lasting ones to revisit in the seasonal round. This is guess work. The first permanent dwellings of

10a

10b

which excavation has unearthed remains are from Iraq and Palestine. They are clay huts much like some still in use near the Nile. In the Old World, they have been found both in the Middle East and in Egypt. Drought may have helped to preserve from destruction the sites on which they stood. Buildings of less durable materials, such as tropical shanties of bamboo poles thatched and screened with palm leaves, would be likely to leave little, if any, trace to posterity. We therefore know little about how real estate development really began.

10. Contemporary reed hut building
(a) basic structure (*opposite*)
(b) completed building

One of the few things we can be sure about is that caves and the cave-men of the comic strip had very little to do with it. What we know about the means of earning a living in the Old Stone Age proves that permanent cave life was never possible, except perhaps to mammoth-trapping Neanderthalers; and what the work of the cave artists of the Old Stone Age can teach us makes it most unlikely that the hairy-chested cave he-man used his club to elope with a reluctant and submissive bride.

Statuettes from the deep cave recesses reserved for ritual practices are of females with very prominent breasts. They are remarkably like clay ones found in the earliest Old World village dwellings. These anticipate frescoes associated in the dawn of civilization with the veneration of fertility and worship of a Mother Goddess. In Oceania, there are still living people who have not grasped that the male has an essential role in procreation. Owing to the long period separating human conception from birth of a child and the fact that human beings, like their nearest relatives apes and monkeys, have no special mating season, as have sheep, it is by no means obvious that he has. So it is likely that the hunters of the cave pictures did not recognize fatherhood as such.

To them, presumably, only the role of the female in the production of a new human being was self evident and worthy of respect. One and all they were children of the Great Mother who is older than all the male gods. The cave male may well have been a meek little fellow in the presence of his mother, aunts and sisters.

4 The domestication of man

To spread far beyond the tropics, our ancestors had to do more than make their own climate. They had also to learn to stock the larder in regions where food of one sort was available in some places during one part of the year and food of another available only at a different season. Wandering in search of small game, edible roots, berries, nuts and eggs of wild birds, they learned the hard way what was abundant here or there in one or another season.

A small family group of a dozen, or at most a score, might support themselves for many years in the same region, returning repeatedly to camp at the same sites. As their numbers grew, they would inevitably exhaust the supply of food sufficient for their needs. Hunger would then force small groups to wander farther afield. Except in so far as they could improve their hunting weapons, could fashion convenient containers both for bringing back what they had gathered to the camp fire and for cooking it, or could make better clothing to protect their bodies, they had little incentive to inventiveness.

A new impetus to invent could come about only when they were able to store a food surplus in settled surroundings. The creation of such a surplus signifies that they entered into a new companionship with other species of living creatures. Instead of precariously hunting game, some learned to tend flocks and herds. Instead of depending on day-to-day supply of plant food in season, some learned to cultivate crops. Either way, the learning process cannot have been sudden.

It is not likely that herdsmanship and crop production originated in one place. More probably migration brought into contact tribes with different lessons learned in different territories. Intermarriage would then give them

44   Beginnings and Blunders: Before Science Began

1. Bushman painting of rhinoceros hunt with dog

opportunities to pool experience. When we probe into the beginnings of sheep, goat and cattle farming, of root crop production and of sowing grain to harvest, we need not therefore picture one or other as an earlier accomplishment. What came before another in one region may have come about after it in a different setting.

In many respects, the course of events which bridge the gap between the Old Stone Age hunter and the first city dwellers was very different in America and in the Old World. Not least of such differences is the fact that the New Stone Age villagers of the New World relied almost wholly on plant life for the food surplus which made possible a more settled existence. In coastal regions or near large rivers, they might supplement crop production with fishing. Except possibly in the mountainous parts of tropical South America, where three small species of camels exist in the wild, the first Amerindian farmers kept neither flocks nor herds.

It is customary to speak of the new companionship between Man and other living creatures characteristic of the Neolithic Age in the Old World as the *domestication* of animals and plants. It would be nearer the mark to speak of how animals and plants conspired to domesticate Man. However, this would be only a half truth. We do not necessarily mean the same thing when we apply the word *domesticated* to animals as when we apply it to plants. As applied to each, it signifies a species which exists mainly, if not exclusively, in a man-made environment and propagates its kind therein. There the similarity ends.

When we speak of domesticated animals we do not imply that man plays an active part in their propagation. When we speak of domesticated plants, we refer to species propagated within a man-made environment by human intervention such as scattering seed or imbedding cuttings in the soil. The second use of the term implies exercise of human curiosity, experience and investigation. As we shall now see the alternative does not necessarily do so.

Whether we ask how men first kept flocks and herds or how they learned to harvest grain they themselves had scattered, relics unearthed by excavation can do little to satisfy our curiosity. Unless we make use of information drawn from different branches of biological science, they can tell us next to nothing. Before we start to dig, let us take stock of what the geographical distribution of plant and animal species can teach us. Two examples are enough to remind us how much we can learn from it. We know that tobacco grows wild only in the New World. If we knew nothing about Sir Walter Raleigh, we could therefore conclude that nobody smoked tobacco in Britain before there was trade between America and the Old World. Potatoes are native only in the Andes. So there could have been no

Irish Potato Famine before the Spanish conquest of Peru, Ecuador and Chile.

Similarly, we do not need to rely on fossil remains to testify that Man's association with the dog extends far back into the twilight of our species. We can reach the same conclusion by studying a world map exhibiting the present distribution of animal life. In every habitable territory of our planet, Man and the dog live together in peaceful coexistence.

With Man, we find the dog in the Arctic wastes. With Man, we find the dog in Australia, where there are no other native placentals, i.e. mammals other than pouched ones such as kangaroos. With Man, we also find the dog in New Zealand where the only true native vertebrate fully adapted to life on land is a lizard elsewhere extinct since the Coal Age. Alike in Australia and in New Zealand, in Alaska, in Greenland and in Tierra del Fuego, both Man and dogs of a sort are interlopers. From the Equator to the Poles, we find Man where we find the dog and *vice versa*.

Let us now look at this long territory-sharing association in the light of what the study of animal behaviour can teach us. Naturalists record that some contemporary wild dogs will follow larger carnivorous animals, such as lions and hyenas, to get their share in the kill. Such a relationship is one which biologists call *commensal*, meaning literally *sharing the same table*. A picturesque example is the squatter sea anemone on the borrowed whelk shell which houses a hermit crab. Like the dog, the anemone feeds on the crumbs from the rich man's meal.

That our remote ancestors in times of severe shortage of food ate the dogs which accompanied them in the chase or hung around their camp sites is not inconsistent

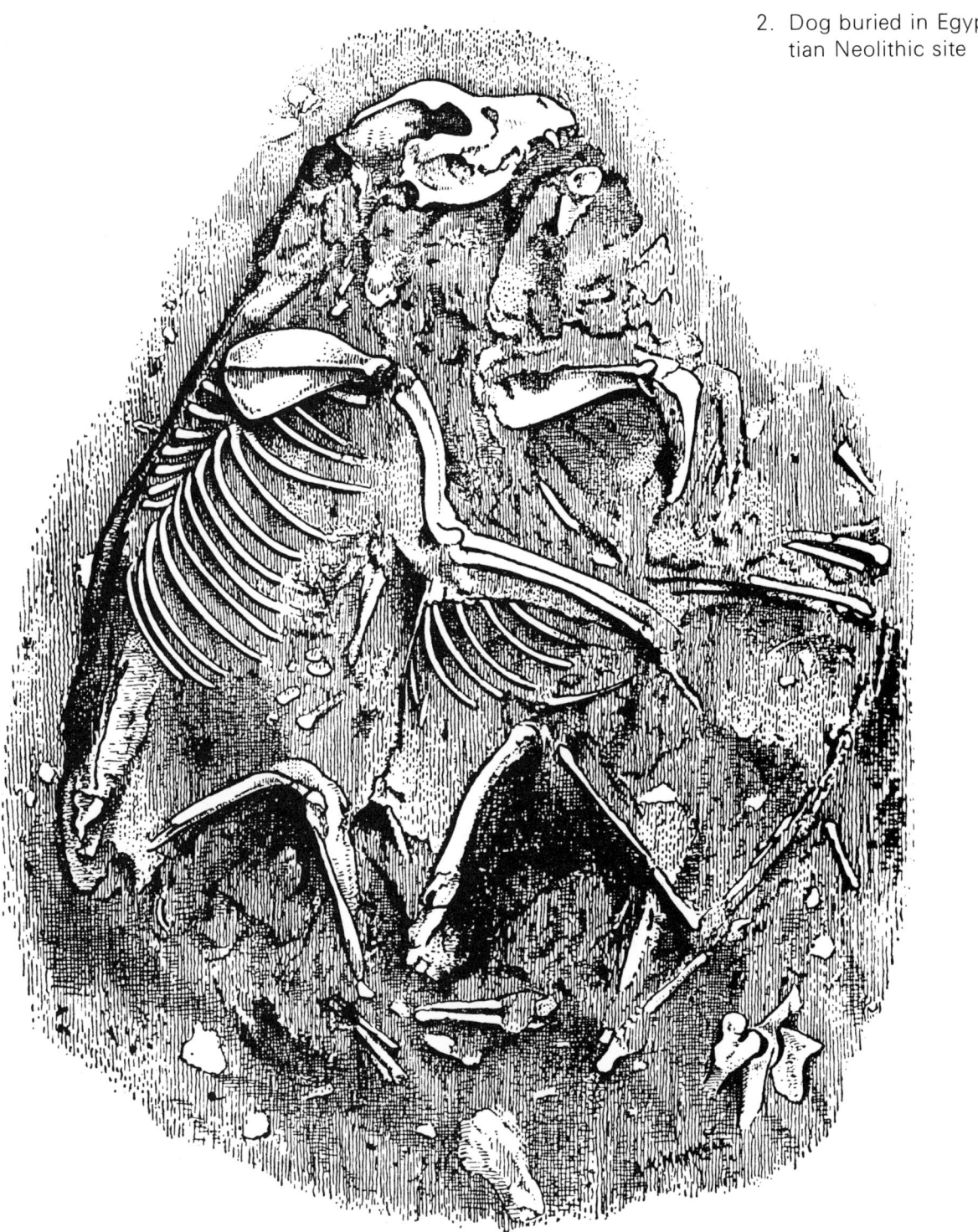

2. Dog buried in Egyptian Neolithic site

with such a commensal relationship. From the nature of their common taste for meat, it was one in which there was give and take. By harassing the beasts Man hunted, his canine partner could make them more vulnerable to the spear or bow of the marksman; and in course of time, the contribution of the dog to the spoils of the chase was to yield bigger dividends. Before we ask what these were, let us see what clues to the association of Man and the dog we can gain from genetics, the science of animal and plant breeding.

First, let us be clear about what we mean by dogs. When one here speaks of wild dogs, one means any member of the dog family called by zoologists the *Canidae*. Within it, they place several species popularly named wolves and jackals. Since domestic dogs of one sort or another interbreed with both jackals and wolves, it is misleading to ask whether the dog is a domesticated wolf or a domesticated jackal. The answer is both and neither.

Wild *Canidae* occur in Europe, Asia, Africa and America. If our picture of the commensal relationship between the dog and the Old Stone Age hunters is correct, we must assume that different wild dogs in different places attached themselves to Man as a competitor for, but none the less a partner in, division of the spoils. Where there were other native dogs in a newly invaded territory they would interbreed to beget new varieties. As men ranged farther afield, there would be greater scope for new combinations of the *genes* which determine the distinctive characteristics of different breeds.

Thus the great variety of breeds of dogs, as of breeds of other so-called domesticated animals, has probably come about by an unwitting process nowadays deliberately applied against a background of newly gained scientific knowledge. Scientific animal and plant breeding does not rely on selection from pedigree material, i.e. inbred stocks of which the hereditary make-up is comparatively homogeneous. More and more, it relies on selecting from hybrids of diverse parentage, including, where possible, wild geographical varieties.

If, as seems certain, dogs accompanied our nomadic ancestors for many thousands of years before men settled down to village life, the fact that there were already widely different breeds of domestic dogs long before the written record begins is therefore intelligible. Pieces of a jigsaw puzzle from diverse sources of biological information thus disclose a picture very different from the traditional belief that Man deliberately tamed the dog. If the commensal partnership between them was a necessary prelude to herdsmanship, it would be nearer the truth to say that the dog tamed Man.

One explanation of the origins of domesticated flocks and herds is that tribes who had begun to combine hunting with a crude form of cereal crop production encouraged intruders to linger near their temporary settlements in search of waste corn or other odds and ends. Those who take this possibility seriously suggest that their human hosts made pets of the young animals. Such a fairy tale does more credit to the kindliness than to the credulity of those who entertain it. That Man became a shepherd or a cowherd by first starting a free-for-all pet shop is not very probable. There is a far more likely interpretation of how he may have blundered into tending flocks and herds. It takes stock of a characteristic of dog behaviour in the presence of gregarious ungulates, i.e. hoofed mammals which normally associate in flocks and herds, such as deer, some species of wild sheep and wild cattle.

A town-bred terrier or labrador brought to the country and confronted for the first time with a field in which sheep are grazing will round them up into a compact bunch. When white farmers invade his territory, a dingo, the Australian dog whose ancestors have never seen a cow or a bullock, will round up cattle in the same way. Where Old Stone Age Man invaded a territory which provided grazing for gregarious ungulates, in particular sheep, goats, or cattle, his hunting partner must therefore have been an even greater asset than elsewhere.

If the territory where they abound has ravines with blind alleys, his commensal needs no bidding to round up a flock or herd in a situation from which there is no escape. When this happens, it would be folly to kill off the whole herd while the dog can mount guard over its one way of retreat. Willy nilly, the hunter finds himself with a surplus. Metaphorically speaking, he now has meat in cold storage for use as required. In short, he has blundered into the beginnings of herdsmanship.

The advantage to the hunter was only temporary. He had still to learn to lead his flocks or herds into new pastures when little grass remained at the site of their capture. Many generations may have elapsed before he did so; but he had a willing partner. It calls for no great effort of imagination to figure how men with dogs to help them gradually acquired the habit of driving their captives to new ground for grazing.

We may thus think of the commensal association of Man and the dog as a curtain-raiser to a food surplus which could free our ancestors from the daily threat of hunger in climates where pasture of a sort was available throughout the seasons. However, the advantage of the association could not last for ever. Even where the herdsman could rely on pasture all the year round, large flocks or herds would eventually exhaust its soil. In colder regions where outdoor maintenance of flocks and herds was impracticable during the winter months, the shepherd or cowherd could settle down to village life with more ample prospects of leisure only when he was also able to store food of a different sort in a new way. In short, he had to learn to cultivate the soil. It may be that his teachers were grain growers who were not themselves herdsmen.

Such few skeletal remains as we have at our disposal to date show that the earliest captive ungulates were what we may call with equal propriety sheep or goats. Even today, the distinction is not everywhere clear-cut. On a Sicilian hillside or on the outskirts of an African village, one may still see mixed herds of which some individuals are recognizably goats, others recognizably sheep and most of them unmistakably neither one nor the other. When and where herdsmanship began there may therefore have been little, if any, difference between the two and there may well have been crossbreeding among the most different.

There are still several wild, so-called geographical species, of either; and several may have contributed to the ancestry of the highly specialized breeds now primarily kept, some for wool shearing, some for slaughter or even for milk production, as are sheep on the Frisian Islands and in Israel today. Where Man led his flocks far afield into territories offering new opportunities for interbreeding with local varieties of stocks different from their own, the story of the dog would repeat itself. New breeds would make their appearance, initially without human anticipation of the outcome.

It is equally likely that crossing between different local species of cattle has contributed to the make-up of our domestic breeds. Among cattle now bred for milk production zoologists recognize three distinct species with different surnames. The ancestor of the earliest European herds has no surviving wild representative. The other two are native to India. Of the latter one is the water buffalo, which has been introduced into Egypt and Hungary. The other is the humped *zebu*, now bred in equatorial Africa. Crossing between different species gives rise to fertile hybrids. It is now playing a part in scientific cattle breeding for tropical territories.

The approach to domestication which we have been exploring thus far is one which biologists call *ecological*. Just as *domestic* and *domicile* both come from the Latin word for a house, *ecology* and *economy* come from the corresponding Greek word, though used in a somewhat more metaphorical way. Ecology deals with how particular species associate in a comparatively stable way as housemates. For instance, we expect to find only certain plants and animals in a pinewood, on a moor or in a rockpool. We do so because the plant and animal species we find in one or another make much the same contribution to the total population in successive generations. Ecology is therefore the study of balanced animal and plant populations.

We may speak of a particular association so defined as an ecological system, and the story of domestication of animals or plants as that of the growth of a *man-centred* ecological system. It is now in large measure world-wide but still has local idiosyncrasies. Elephants, camels, reindeer and vultures have a very limited range as human housemates, though each has had, or still has, considerable influence on human history: elephants in the Persian campaigns of antiquity, camels in the mediaeval Arab slave trade, reindeer in the Arctic regions of Asia and Europe.

The inclusion of the vulture in the last list of animals may have pulled up the reader with a jerk. It is not a joke. If one visits the open air food markets of West Africa, even those of sizable towns, one may watch the vultures hovering above throughout the working day. As they hover, flies buzz to and fro between the cassava or banana *foufou* (paste) and fish heads or edible giant land snails which putrefy quickly under tropical sun rays. Within a few minutes after the last of the mammy traders has packed up, the expectant birds descend. In a short time, they have cleared the site of any trace of organic matter which would otherwise become a breeding ground for fly-borne diseases.

As a human commensal, the vulture therefore pays for its keep. In regions where there are no sanitary authorities, it performs a useful service as a scavenger. It is probably as a scavenger that the pig became incorporated in the human ecological system. It remains as such in the neighbourhood of villages in some parts of tropical Oceania. In earlier times, both pigs and poultry may have visited the litter of abandoned camp sites for what pickings they could collect. In the neighbourhood of more permanent settlements, their human hosts probably became more tolerant to their presence as village life took shape. The cat and the horse were relatively latecomers.

When storage of grain began, it invited mice as unwanted housemates and it is likely that villagers welcomed the intrusion of the cat as a mouse catcher. How the human ecological system recruited man's two most widespread beasts of burden is still a mystery on which neither biology nor excavation as yet throws any light. The ass was a pack animal when city life was still a novelty in Egypt and in the Middle East. That was long before horses reached them. Possibly its association with Man started through its use as a source of milk. Horsemanship began in the plains of central Asia. We know warlike tribes spread it as a means of marauding their neighbours' territory. Of how they learned to break in their steeds, we know as yet nothing whatever.

5 Crops and containers

The word civilization comes from the Latin word *civis* meaning a citizen. By the beginning of civilization, one therefore means the beginning of city life. Before there could be cities, there had to be smaller communities with fixed dwelling-places. Life in such villages could leave traces to posterity only when and where the dwellings were of durable materials.

That their occupants could build homes able to outlast their own lifetime and that of their children was possible only when communities had adequate supplies of food close at hand throughout the year. In the Old World some small groups of families may have first blundered into keeping flocks of sheep or goats and others into herding cattle. If so, they would have to rely on an exclusively meat diet unless they either kept on the move or could store plant foods. Without a store of plant food, they would not be able to insure survival of their livestock, where winters were severe and abundant pasture seasonal.

Nor would an exclusive diet of meat, sun-dried or salted for winter storage, have sufficed for Man's own needs. It would be deficient in constituents essential to health. So the discovery of how to store plant food was a milestone in the life of mankind; and suitable plant food had to meet two requirements other than ease of storage: adequate protein content and quick growth. Tree-borne nuts meet the first requirement but not the second. Root crops, such as abound in the tropics, meet the second but not the first.

Cereals are satisfactory in both ways. When our forefathers learned to sow grain and to harvest it, several family groups in nearby dwellings could enjoy a balanced and sufficient diet. Thus village life became possible. Excavation has unearthed its first traces in the Old World

1. Wild species of wheat
   (a) diploid wheat

at Jarmo in Iraq and near Jericho in Palestine. Archaeologists have there found remains of lambs or kids with grains of wheat and barley on the sites of human dwellings. We have thus unmistakable evidence that mixed farming and village life had by then begun. In the Middle East, farmers of the New Stone Age already had flocks and already harvested grain before 7000 B.C.

To say this does not necessarily mean that farming of either sort began there. Nor does it mean that we can confidently explain how mixed farming began. That the earliest traces of farming so far excavated in the Old World are those of villagers keeping flocks as well as growing grain does not prove that either innovation came before the other. Excavations elsewhere may eventually unearth earlier traces of village life than those of the Middle East. As we have seen in Chapter 4, the earliest villagers of the New Stone Age in America were not herdsmen. Before the coming of European cavalry armed with muskets, so-called domestication of animals played a very minor role in the civilizations which flourished in the New World. Indeed, it did so only in the Inca empire of Peru and neighbouring territory.

In America, where maize was the only cereal crop grown before the Spanish Conquistadors arrived, discovery of how to cultivate it was certainly independent of discovery of how to cultivate other cereals in the Old World, where there was no maize before the time of Columbus. The earliest villagers of America cultivated beans and squashes, i.e. pumpkins and their like, before they cultivated the so-called Indian Corn.

In different parts of the Old World, different cereal crops have come into use; and we have no reason to believe that the know-how of all the cultivators is traceable to one source. It seems that the first Europeans to grow oats and rye already had experience of growing wheat and barley. It also seems that the Chinese, and likely enough the Indians, grew millet long before they started to cultivate rice whose wild ancestor came from the East Indies. It is by no means so likely that the first to cultivate millet learned how to do so from others who cultivated wheat or barley.

To get into true perspective what our first peep at village life in Iraq and Palestine and Egypt can teach, we need to take a world-wide view of the cultivated cereals and where their wild ancestors live. It is not easy to pin down the wild ancestor of maize. We know more about which cereals were grown on the earliest Old World village sites. In Palestine, Iraq and Egypt, wheat and barley were the earliest cereal crops cultivated. Two wild species of each exist. Microscopic study shows that they are undoubtedly ancestral to the cultivated forms, and both pairs of wild species grow only in the Middle East. It is therefore likely that the practice of cultivating wheat and barley spread from Palestine into Egypt at a very early date.

What excavation has unearthed in the relatively small region where the ancestors of wheat and barley grow wild would amount to more if wheat and barley were the only cereals cultivated by Man. They are not. From time immemorial in the hotter parts of the Old World, people separated by thousands of miles have sown and harvested one or other of several species called millets or sorghum grasses where such species also grow wild.

With all other edible seed crops, except buckwheat introduced into the U.S.A. from its home in Central Asia, botanists place wheat and barley in the grass family (*Graminaceae*). So we may properly speak of their ancestral forms as wild grasses. Whereas wheat, barley,

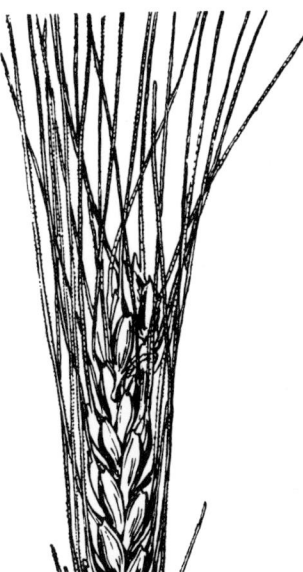

(b) tetraploid wheat

oats and rye are not suitable to cultivation in very hot countries, other cereal crops and their wild forms are widespread in the tropics. Of these, rice is of least interest to our story. Its mode of cultivation involves more continuous labour than that of other seed crops, and it is doubtful whether either the Indians or the Chinese would have learned the trick of growing it, if not already skilled in growing millet.

Millet and sorghum are popular names for half a dozen wild grasses cultivated in the tropical belt of the Old World, and some of them can propagate successfully in warmer regions north of it. Sorghum, known as guinea corn in West Africa, grows abundantly as a roadside weed in Ghana. Ghanians cultivate or collect it to malt for brewing a local beer. Elsewhere in Africa, it yields a bread flour. Millet, as grown for bread-making in India and in tropical Africa, yields the flour for a loaf more agreeable to some palates than wheat bread, now chiefly consumed in western Europe and North America. Just as wheat bread has a less coarse texture than the black (rye) bread of Eastern Europe, millet bread has a texture less coarse than that of wheat bread.

Till there has been much more excavation in regions where edible cereals grow wild, we should keep an open mind about how, when and where cereal crop cultivation began. Meanwhile, it is easier to believe that cultivation of grain crops began in several widely scattered regions where men collected any one of a dozen species of wild grasses for food than that it started once for all in the very small part of the Middle East where wheat and barley still grow wild.

From what we know of communities still very backward in the middle of the last century, we may guess that our nomadic forefathers of the Old Stone Age circulated in search of food between the same camping sites for several cycles of the seasons. Where the territory had abundance of wild grasses with edible seeds, they might leave the droppings littered near their camp fire kitchens. If they did so, they would find grain sprouting at the same site when they returned to it. They may thus have learned gradually that it paid to leave a little grain as litter. None the less, it involved an act of foresight to set apart some of their store to sow for harvest on the return journey.

So likely a story leaves us with no answer to three questions. We do not know how Old Stone Age hunters first acquired a taste for the seeds of wild grasses. We do not know what containers they used for collecting or storing them. At a time when they had not learned to make earthenware, we do not know in what vessels they could make a stew or porridge from the grain.

We are, alas, on very thin ice when we ask how our ancestors acquired a taste for wild grass seeds. It is unlikely that they collected them in bulk to eat raw, especially if we assume that cereal cultivation started where wheat and barley grow wild. The earliest wheats were indistinguishable from barley in the way most of us now tell one from the other. They had long prickly awns. So it is difficult to believe that our hunting and hungry forefathers would have plucked them to make an uncooked meal.

Where no wild birds' eggs, no roots nor fruit were in season, men and women haunted by hunger between one successful hunt and another would be ready to try strange diets. Still, it is more likely that they acquired the grass-eating habit by chewing unripe cobs of tropical guinea corn than that they did so by munching the prickly ears of wheat or barley. Even so, the ripe grain

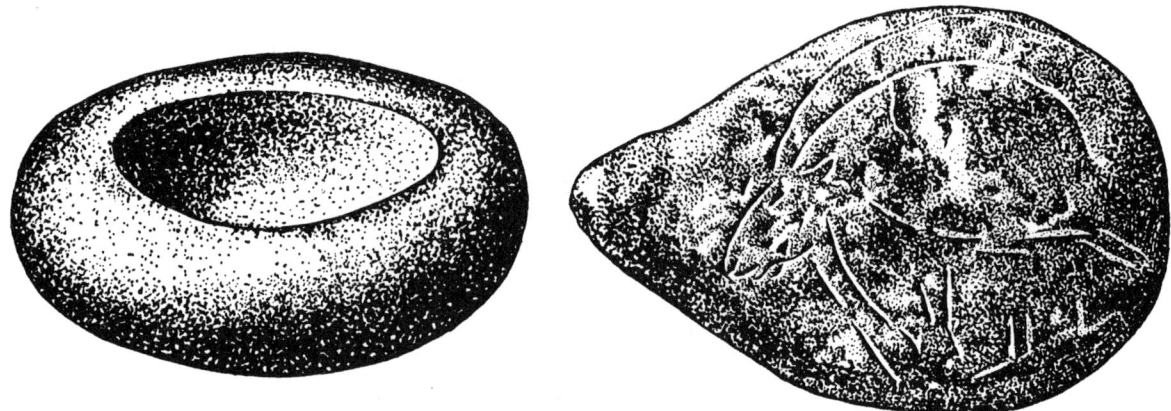

2. Palaeolithic stone lamps

when stored would not be easy on the teeth of the newly weaned or of the old folk.

If we care to speculate about how our Old Stone Age ancestors acquired the taste for a cereal diet, we have no need to be squeamish about their table manners. There is no reason to believe that they were more fastidious than our contemporaries in the northernmost part of the New World. Canadian Eskimos, whose way of life at the beginning of our own century was in some ways much like that of the later Picture Makers of the Old Stone Age, eat the entrails of the caribou, a major source of food for them in the Arctic summer. The partially digested lichens in the contents of the gut supply them with their main source of vegetable food, and hence of vitamin C to protect them against scurvy.

Let us suppose that the Picture Makers cooked their smaller prey without disembowelling them. Among the sizable birds most easy to catch would be those of the pheasant family. This includes partridges, grouse, quails, turkeys, the African guinea fowl and the Asiatic bantam jungle fowls ancestral to our own domestic breeds. Their distribution is widespread throughout the warmer parts of both the Old and the New World. Though they will devour other forms of plant food most of them eat seeds, especially of wild grasses where available. That they do so, accounts for their intrusion into what we have now learned to call the human-ecological system.

After a meal of wild grass seeds such as guinea corn, the crop of such birds serves as a container for grain sifted from chaff and softened by the juices it secretes. If eaten with the rest of the roast, its contents would not necessarily have been distasteful to the Old Stone Age hunter and trapper of small game. Like the stuffing of a Thanksgiving or Christmas turkey, they may have given the meal extra relish.

There is another possibility. In parts of tropical Africa where millet is the cereal cultivated, the only use made of the wild guinea corn is to brew a beer. Where the Africans make palm wine, they do not skim off the yeast. Seemingly, they find it palatable. Possibly, therefore, the chance discovery that grain soaked in water ferments gave Old Stone Age Man the first impetus to collect grain. If so, the first intention of adding yeast to flour need not have been to make the dough rise. Maybe the villagers of the New Stone Age used it to flavour the dish.

If we cannot be certain about what circumstances prompted our hunting ancestors to rely on cereals as a major source of food and to make the grain more or less palatable and digestible, we can have no doubts about two consequences of doing so. When they first started to collect wild grasses they may have brought home the stalks by hand; but when they started to store it they would need containers of some sort. When they started to harvest it they would also need a cutting tool equal to the task.

Whatever containers our ancestors used before they took to making pottery must have been of far less durable material, otherwise it would leave no trace except in exceptionally dry situations unfavourable to human survival. This is equally true of bags made by sewing animal hides and baskets made by plaiting reeds or twigs. So we cannot put a date to when people first began to make them.

The earliest traces of the art of basketry in the Old World are also the earliest remains of a container certainly used for storage of grain. They are from an Egyptian village site at some date earlier than 4000 B.C. Basket work may well have been of far greater antiquity. Since basket making occurs both in the New and in the Old World, it probably started long before we first find traces of it in the New Stone Age village sites.

Since some of the villagers of the Middle East had stone bowls before they made pottery, such vessels may have been among the earliest cooking utensils. The earliest container used for making a stew may have been a leather bag brought to the boil by immersion of pebbles previously heated in the camp fire. Bedouins of North Africa use such bags as containers for milk, and the later Picture Makers had bone needles suitable for stitching animal hides. When they wandered far from springs, lakes or rivers, they would need some such receptacle to carry drinking water.

We know more about the earliest cutting tool for harvesting a cereal crop than about containers used for gathering grain or cooking it. At least as early as 8000 B.C. cultivators of grain crops in the Old World had contrived to make a sickle by fitting thin flint blades into a groove cut lengthwise on an antler or thigh bone. Experience of preparing the soil for seed growing after the harvest taught the apprentice farmers of the New Stone Age the need to improve tools of greater antiquity.

At a primitive level of farming, cultivation of any sort

is possible only by turning the clods to give a fair chance to seedlings. Any of the digging tools of the first farmers may have been adaptations of cruder devices used for getting at edible roots of the sort abundant in the tropics. Foremost among such was the stone-weighted digging stick still used by the Bushmen of South West Africa. Shoulder blades at first served as shovels, and antlers as hoes before polished stone heads with wooden handles replaced them.

The use of the digging stick raises the question: did root crop cultivation precede that of cereals, and if so where? No answer to the question makes sense unless we realize that the term root crop has two different meanings. The edible part of turnips, beets, carrots, radishes and parsnips, i.e. that of all root crops native to the cooler parts of the Old World, is a tap root, and is useless when it reaches a certain size. That of potatoes, yams, cassava (tapioca) is a cluster of tubers, i.e. swollen underground stems. The cultivator can retain any one tuber of such a cluster to propagate a new plant without recourse to sowing seed. Thus root crop production in the tropics of the Old World does not rely on seed propagation. To cultivate yams one need only leave some of the tubers in the ground. To cultivate tapioca one need only use the woody stem as a cutting.

The two edible root crops now propagated without recourse to seed in temperate parts of the Old World are comparatively recent immigrants from the New World. One is the potato, the other is the Jerusalem Artichoke. Thus root crop cultivation could not have started before that of cereals in the cooler parts of the Old World, where the farmer has to rely on seed sowing to propagate the only native edible roots. In Europe, even as recently as four centuries ago, root production was not widespread.

In the tropics, it may well have started before people learned to sow grain.

Our account of plant domestication would be incomplete if it took no stock of a few species which the early farmers of the Middle East learned to cultivate along with cereals before 5000 B.C. The list includes legumes (peas, beans, lentils), dates, apples, olives and flax. The last two are rich in fat. Flax itself had another more important use, as we shall see when we take a look at the household crafts of the first farmers.

The cultivation of legumes at so early a date in the Old World tempts one to speculate about what might have happened if human beings had cultivated them before they developed the taste for a cereal diet. The seeds of legumes store well. Since the nitrogen-fixing bacteria of their root nodules make them able to utilize atmospheric nitrogen, they grow in unmanured soil, i.e. soil with a low nitrate content. They are rich in protein and hence suitable to supplement a diet of root crops, such as yams and cassava (tapioca), which flourish in the tropics.

One tropical legume is very rich in fat. We speak of it in different parts of the English-speaking world as the pea nut, monkey nut, manilla nut, earth nut or ground nut. In parts of tropical Africa where the dreaded *tsetse* fly has hitherto made stock-breeding and sheep rearing impossible, it offers the human population the prospect of a more balanced diet and freedom from dietetic deficiency disease. The history of their continent might therefore have been very different if Africans had learned to cultivate it in the remote past. This could not happen. Like cocoa, maize, tobacco and the potato, the peanut is a native of the New World. It could not reach Africa or Asia *via* Europe until after the time of Columbus.

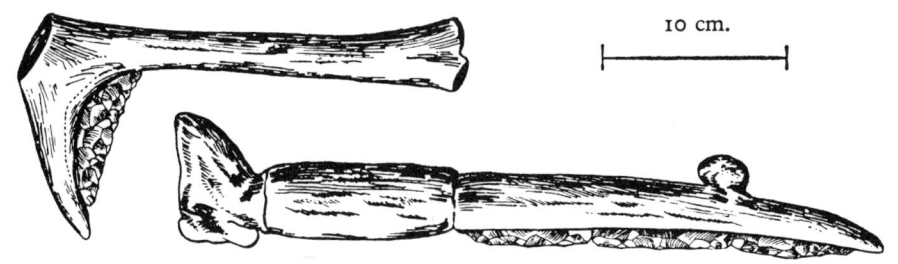

3. Mesolithic and Neolithic sickles

In taking to cereals, the New World had its own problems. Maize was the only cereal cultivated in America before the Spanish Conquest. The build-up of its protein molecule is deficient in one brick essential to human health, and it does not contain the high content of B vitamins found in wheat and barley. One biologist has suggested that too great reliance on maize as the staple food helped to shorten the duration of the earliest American (Mayan) civilization whose remains we now find in the jungles of Central America. However, the Amerindians had insured themselves against too great reliance on maize and had thereby anticipated one immensely important agricultural discovery of modern times long before the Europeans came on the scene.

Not till the eighteenth century of the Christian era did Europeans hit on the practice of alternately sowing a leguminous crop, such as clover, alfalfa, etc for cattle or horse fodder and a cereal or root crop for human consumption. Barely a hundred years have passed since we learned why it is a good thing to do so, i.e. to understand that the root nodules of the legumes make use of atmospheric nitrogen to enrich the soil. Centuries, and it may be several thousand years, before Europeans began to rotate leguminous and other crops, Amerindians had learned to do much the same thing. They made a practice of planting beans in the same plot side by side with their maize plants and squashes. We shall see that this gives us a clue to one difference between the local conditions in which city life of a sort began in the two hemispheres.

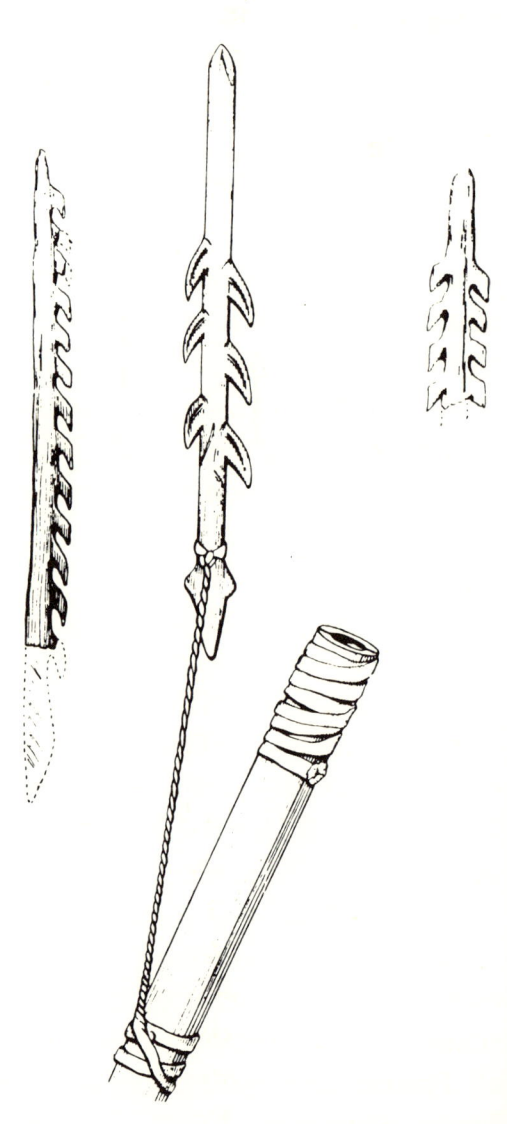

1. Old Stone Age harpoons

6 Fellers of trees, fishermen, trappers and traders

We have now seen that the change from a wandering to a settled way of life did not follow the same course in the New World as in the Old. None the less, one outcome of cereal crop production was the same in both. Release from a precarious daily search for food gave at least to older members of the tribe and to some of the womenfolk more free time. They could now practise new crafts, among which were pottery making and weaving from spun fibre. Both arts began in America little, if at all, later than in the Old World.

In the Old World, the main impetus to village craftsmanship came from beyond the tropical zone where communities suffered neither extreme heat nor severe cold. There, the earliest Old World civilizations took shape when village life was spreading into cooler northern regions destined to become the birthplace of modern technology.

The native civilizations of the New World never extended beyond the fringe of the tropical belt. They developed at a much later date than those of Iraq, Egypt, India and China, and contributed little, if anything, to the world-wide civilization of today. For us therefore, spread of village life, as a prelude to city life, in the northern half of the Old World, and more especially in Europe, is of unique interest and our main concern in this chapter.

Before mixed farming began in the cooler north, local conditions had already been favourable to settlements of small family groups. The ice cap which had embraced all France and part of Spain when peopled by the Old Stone Age cave artists was still receding. Once-fertile parts of Africa and Asia were becoming deserts. Both changing distribution of fertile soil and more bearable weather thus gave an impetus to migration northwards.

In Middle Europe, new climatic conditions immediately following the recession of the polar ice cap favoured the growth of vast forests, and while this was happening, the rising sea level of the coastal region of the continent was exposing longer tracts of seashore. Thus late Old Stone Age migrants trekking north into increasingly habitable regions encountered new resources of food supply. Some of these paved a way to a mode of life still pursued by the Eskimos of Greenland and the New World.

Some migrants became less dependent on hunting land animals. Where the trek took them to the sea coast, they could subsist largely on fish, shellfish and seabirds. Seals, which they captured with the harpoon, supplied them with meat, fat for crude stone lamps and much-needed hides for clothing. Where the trek took them into dense forest territory, they followed the course of wide rivers, such as the Danube, the Dnieper and the Don, teeming with fish throughout the year. In such situations, voyagers could form settlements by the seashore or in clearings of the forest near the river bank.

Clearing the forest for space in which to settle was practicable only after addition of a new device for their tool kit. They had to make a heavy stone axehead with a sharp cutting edge and attach it to a wooden handle. Partly because the rivers were the highways to travel through dense forests with no roads, partly because the migrants were so largely dependent on a fish diet, the northward movement encouraged the perfection of two other devices. One was fishing tackle. The other was the boat.

We have seen that the later Picture Makers had harpoons made of bone or antlers. Their successors had fish hooks of pierced and pointed bone much like those of Eskimos. They also had nets of a sort. Cave pictures make

Fellers of Trees, Fishermen, Trappers and Traders 69

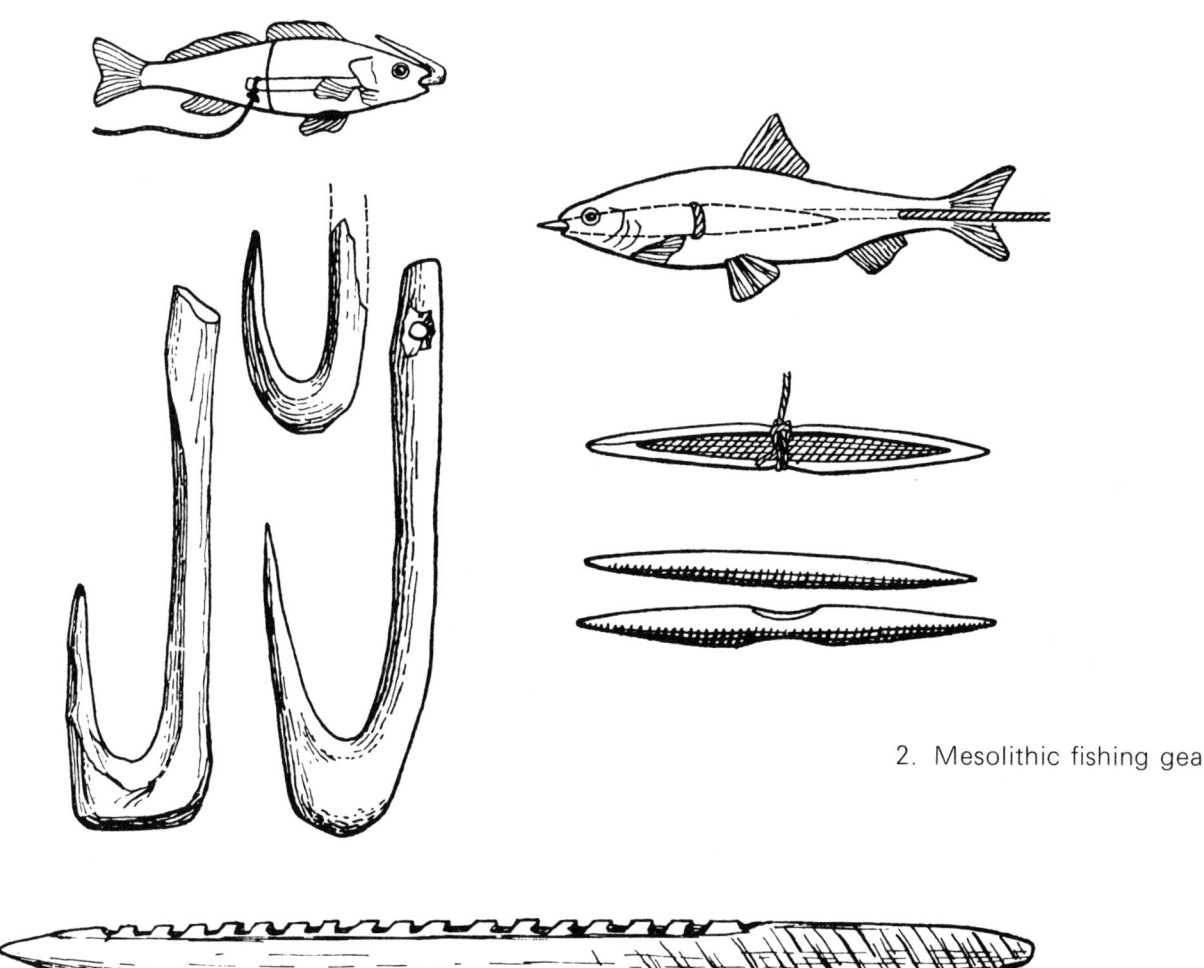

2. Mesolithic fishing gear

it certain that the forerunners of the forest dwellers at the end of the Old Stone Age in Europe already used bows and arrows; but no such picture can give us an equally certain answer to the question: what material did they use for the bowstring? Likewise, we have little reliable information about what sort of material inshore fishermen and fellers of forests first used for making nets, lines for fish hooks and cord to attach the stone blade firmly to the wooden handle of the axe.

It is likely that the fishing tackle and hunting gear of

the forest fellers of northern Europe in the last phase of its Old Stone Age were much like those of Eskimos before they came into contact with traders. Before any of our ancestors grew hemp for rope-making, the toughest and most flexible material for binding and knotting was a thong cut from the hide of a mammal such as a seal. But cords of hide are not very durable, so we are not likely to unearth remains of them.

Nets made of such material must have had a wide mesh, and were therefore suitable for catching only large fish when slung across a river. The forest settlers may have used floating logs to support them. If so, we have a clue to one way in which they first navigated. Once they got used to sitting astride a floating tree trunk, it was a short step to making a canoe.

In some parts of South America, the riverside Amerindian villagers still make one by first burning a deep cavity lengthwise in a large tree trunk. They then finish the job with what cutting tools they use for tree felling. Even before some forest fishermen hit on the device of charring a cavity in which to sit, others may have noticed that swinging a small branch sideways could propel a floating log to carry a man more quickly downstream. From so simple an observation, the navigator, already accustomed to tree felling and to making axe handles, would find out how to fashion a paddle.

The waterways of northern Europe and Asia cut across barriers of forest and penetrated between mountain ranges. So the construction of boats enormously speeded northward migration of tribes which subsisted largely by fishing. One consequence was spread of population into territory as remote from man's tropical homeland as the extreme north-east of Asia. Thence, trekkers crossed to Alaska *via* the chain of islands called the Aleutians.

Fellers of Trees, Fishermen, Trappers and Traders 71

Those who did so cut themselves off from the cultures of the Old World by venturing into regions whose animals and plants were unlike any known to their ancestors. Inevitably, therefore, the sequence of events leading to village life in the New World was different from that in the Old.

Canoes, first made of tree trunks, and later of hides stretched over a wooden frame, increased man's mobility in another way. At least in Europe, traffic was not all northwards into uninhabited territory. Once the forest fishermen and dwellers near the seashore had learned to make boats of a sort, and had mastered the art of navigating in rapid rivers or along the coastal fringe, hitherto isolated communities which had subsisted largely by fishing came into contact with people who had other folk ways. Fishermen became middlemen. As middlemen, they promoted a mingling of skills.

Where their journeys brought them into contact with farming communities, they shared a new role with trappers who could thread their way through forests. Folk who could travel afar in boats, and trappers who could blaze a trail through forests, could now trade by barter with settlements having no other means of communication. As demand grew for what they could offer in exchange for grain or pottery, there was a new incentive to make better boats. They could then take longer

3. Stone Age rock picture of boat

cruises on wider waterways to sites where flourishing communities of farmers had a surplus of food to offer in exchange for foreign materials to develop the household arts which a food surplus could encourage. By their voyages, the know-how for cultivation of crops therefore spread northwards. Log cabin fishing settlements became villages occupied by farmers.

Where the villagers had wholly abandoned hunting, the boatmen and trappers had much to offer in exchange for their surplus of grain, meat or hides of goats. They could thus venture on longer voyages to stock the larder in preparation for expeditions to explore unfamiliar creeks or territories. Their own settlements could accommodate larger clans. They took on a more permanent character with more substantial shanties for dwelling.

Till our own century began, trappers who live by trading fur have been the only line of communication between the Eskimos of the Canadian Arctic and metropolitan Montreal. A comparable situation existed as the Neolithic way of life penetrated farther north in Europe and Asia. When a primitive level of farming reached Britain about 3000 B.C., trappers who knew the trails through the forest and fisherfolk familiar with rivers or coast-line were the only link between people who established settled agriculture on Salisbury Plain, in Cornwall and in Wales.

Excavations near the site of the great British Sun Temple of Stonehenge built about 1800 B.C., have shown that flint implements used by the New Stone Age farmers of one part of Britain could not have come from their own locality. For making polished stone tools, the earliest farmers in Britain and elsewhere on the mainland used rock from distant territories, separated from their settlements by dense forest country and with only navigable rivers to provide a route for heavy cargoes. For what the tillers of the soil could offer, such folk could barter stone of superior quality for making axe heads, sickle blades and hoes.

What we have learned since aerial photography stimulated renewed interest in the habits of people, now named the *Wessex Culture*, near Stonehenge has shown us that boats capable of carrying vast masses of rock for its construction plied between the west coast of Wales and the Bristol Channel for the benefit of the builders. We know this because stone used to make part of the great Sun Temple is of a sort found only in Wales, more than two hundred miles away.

To a large extent, the first farming settlements in northern Europe developed their own way of life only by remaining in friendly contact with the fisher boat-men and forest trapper. The first farmers of the northern fringe had much in common with them. They themselves came of a stock which had settled on the land while still dependent on raw materials which their forbears procured from the quarry they hunted. They must therefore have exercised the same skills to fell trees for making wooden ploughs and to cure hides for apparel or for bags sewn together with bone needles.

Flint of a quality best fitted for making smooth stone hammer heads, axes or blades for mattocks and hoes is not often abundant near soil most suitable for pasture or crop cultivation. Nor do we usually find in good farming country clay suitable for pottery-making, the right twigs and reeds for making baskets or mats and fibres of a sort best for spinning or weaving. In short, land fit for cultivation in northerly regions can rarely have been near sources of supply on which the tillers of the soil depended for the raw materials of village crafts.

4. Log canoe made by burning out a cavity

Only thus could the itinerant trapper or boatman keep them in touch with their sources. With or without a canoe, the trading trapper thus became the means of passing skills from communities in need of commodities to communities with the commodities but without the skills. In the process, the village necessarily became less self-sufficient.

Raw materials for tools and for household crafts were not the only commodities which itinerant fishermen or vagrant trappers of the forest could offer to the village community. From time immemorial, human beings have valued bodily adornment; and the first farmers were no exception to the rule. As the villagers became more prosperous, there was therefore an expanding market for shells, amber, and coloured stone such as blue lapis lazuli to make necklaces or armlets for ornament or as status symbols. Before metal tools replaced stone ones, the demand for gold and silver had already begun, and for the same reason. The earliest remains of gold ornaments in the south of England came from Ireland at a time when village life in Britain was still in its infancy.

As the volume of trade increased, the way of life of the traders must have become less and less precarious. By 3000 B.C. there were already people accustomed to city life in the fertile river valleys of Egypt and Iraq; but as yet there were few farming settlements in Asia Minor, in the Caucasus, on the shores of the Caspian, in South Europe and in the basin of the Danube. The existence of large concentrations of population with well-stocked granaries farther south offered more and more rewards to traders who had learned not to fear the unknown. Thenceforth, there was a mounting demand in Egypt and in Iraq for metals and many other materials, not all of them useful. There was also a market for hardwood, for spices and for mineral pigments such as ochre to beautify pottery or temple precincts.

As early as 2500 B.C. we learn of merchant colonies from city states in the heart of Asia Minor. The great mineral wealth of their territory gave them, says Sir Leonard Woolley, "an importance quite out of proportion to the cultural level of its inhabitants". In short, the impact of traffic conducted by people who still lived by hunting and fishing, or as part-time subsistence farmers, must have been drastic. There was now need for specialists able to make fine pottery, to sift gravel for gold dust, to mine and refine ores of copper, tin, and later, iron. It was a demand for experts living remote from city life.

Better tools for the trading settlements were doubtless available as wares in exchange for materials valued by the wealthier caste of the city. Knowledge of farming techniques thus spread in the wake of the trader. Only by improving their farming methods could herdsmen with temporary settlements or villagers still new to grain-growing produce a food surplus to support specialist personnel able to exploit effectively resources greatly prized by city folk.

As the New Stone Age way of life spread northward, transition from the boatman-trapper to the trader was neither local nor abrupt; but it gathered momentum. Still dependent on nature for essential raw materials, the first farmers could not withdraw themselves wholly from their itinerant neighbours. Only with their help could they themselves be free to settle where soil and contour of land were most favourable to the practice of agriculture. That barter trade began so early explains much which would otherwise be unintelligible.

The first people who built cities with monumental buildings still able to excite our admiration and wonder were not outstanding for their contribution to the momentous step which mankind made by mastery of metals. Progress in mining and moulding them came from far poorer and far less populous communities separated by long distances from the cradles of civilization. In terms of literacy and the arts of leisure which flourished already in the city states of Egypt and Iraq, the folk who at first contributed most to the making of metal tools and weapons were savages.

Between such communities and the cities, the trader—in some territories still part-time trapper or fisherman—was the middleman. Without him, the pioneers of metallurgy would have had no market and no means of subsistence to ensure the exercise of a newly found skill; but the middleman was not a by-product of

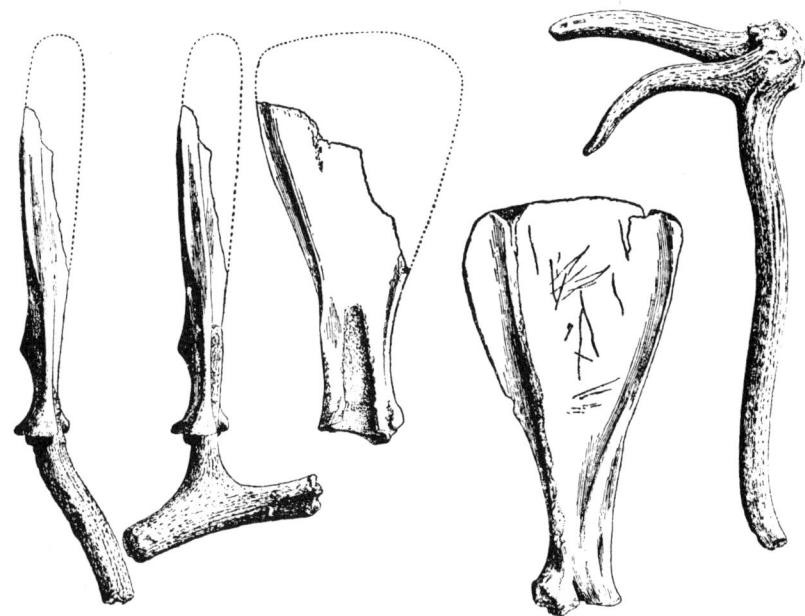

5. Antler picks and scapula shovels

metallurgy. Barter between the New Stone Age farmers and clans with a less settled life had gone on long before the first use of metals. Indeed, we may expect to find traces of trade millennia before the beginnings of metallurgy or the dawn of city life. What was new about trade in the setting of the city temple sites is that it offered novel and richer rewards for a nomadic fringe.

In most of this chapter our concern has been with the spread of settled life northward as a curtain-raiser to the emergence of European civilization. This has put the spotlight on the role of the fisher-boatman as an agent for spreading knowledge of new skills. At a time when the now-barren sites of temples in Mexico were dense forests, that of the trapper and feller of trees may well have been important in the New World. As we shall later recall, the native civilizations of the New World did not come into being on the banks of its great rivers, and there is scant evidence of sea-borne traffic between them. The Inca Empire in Peru, Ecuador and Chile relied on its trunk roads for communication between its several centres of population.

7 Handicrafts among the first farmers

Readers of other books about our remote past may come across a picturesque expression for the beginnings of village life. Some experts speak of it as the Neolithic Revolution. To do so is somewhat misleading. The transition from a more precarious and nomadic to a more secure and settled existence was not an abrupt event like the American or French revolutions. Its character was not identical in different regions and its consequences are not certainly traceable to a single source.

It is not even true to say that village life of a sort began only as the result of cultivating the soil, keeping flocks or both. As we have now seen, settlements of fisher folk along the coastal strip or forest waterways of Northern Europe existed before the occupants learned the art of cultivation from their southern neighbours. We have also seen that the pattern of the most primitive farming in America is very different from the sort associated with cereal cultivation in the Old World. Community life based on a combination of crop production and fishing existed on the west coast of tropical South America more than four thousand years ago. This was long before the villagers began to cultivate maize.

A comparable pattern of community life has long existed in some of the chain of islands between Asia and Australia. The villagers do not cultivate cereals. They subsist partly on breadfruit and root crops (yams and taros), partly on sea fish and clams, with an occasional feast on the pigs which intrude as scavengers in the neighbourhood of human dwellings. How much or how little their way of life owes to the cultures of regions from which they migrated in a very distant past is by no means certain.

By whatever route communities in different regions adopted some sort of farming suitable to local conditions,

Handicrafts among the First Farmers 79

1. Neolithic house in Iraq

one consequence of having created a food surplus was common to all. There was greater leisure with greater specialization to practise such domestic handicrafts as basket work, mat making, pottery, rope making, spinning and weaving. Of such, spinning and weaving did not begin till village life was well established. Since mats would be an encumbrance to people with no fixed dwelling place, we may assume that the same is true of mat making.

Of the village handicrafts mentioned, pottery has an intelligible fascination for the excavator. Being breakable, it calls for frequent replacement. It therefore

2. Basket-lined grain storage pit

accumulates among the litter its users leave near their dwellings. Being resistant to both moisture and heat, its fragments are well nigh indestructible. In short, we are almost certain to find some remains of earthenware vessels at the site of any village or camp, if those who once occupied it knew how to make, or used, containers of baked clay.

Containers of earthenware are immensely more satisfactory than any used before them for cooking, for carrying or for storing water, for drinking and as receptacles for salt. In some regions, the last-named was not a trivial advantage. In a cool climate, we are apt to regard salt merely as a preservative or as a means of giving meat or vegetables a more palatable flavour. In hot countries, its use is important for another reason. It is a necessity of health to replace the mineral loss from excessive sweating.

The invention of pottery was thus an event of no little importance. Its many uses and the world-wide occurrence of the raw material for making it partly explain why it is so widespread even among backward communities. Manufacture involves four processes. The first is to moisten the clay to a suitable consistency for the next stage, i.e. moulding it to the required shape. After shaping, it is necesary to dry it very slowly to avoid cracking in the final stage, i.e. heating in some sort of kiln to the appropriate temperature.

At an early stage in the art of making earthenware pots, the maker may have blundered into the discovery that rotating a basket-work tray with one hand while pressing on the moist clay mass with the other simplified the moulding process; but the earliest villagers to make pottery certainly lacked tools equal to the task of making a serviceable potter's wheel. Before that, the potter used either of two methods, both practised among backward communities today. One is to build up the vessel by superimposing successive rings or spiral coils of moistened clay before drying and cooking it. The other is to encrust with softening clay a frame of basket-work which will burn away when the dried vessel is baked.

The first method recalls one primitive way of making a basket from twisted ropes of grass bound together by stitching. Both ways of making an earthenware vessel therefore suggest that the art of making baskets preceded that of making pottery. There is no doubt that it did so in America; and the fact that the earliest remains of basket work so far found in the Old World are from sites more recent than those from which the first fragments of pottery come does not conclusively disprove the possibility that what happened in the New World also

3. Early basketry from Peru

happened in the Old. Where the materials are available, basket-making is well-nigh universal in both hemispheres. Throughout the tropics, basket work of woven twigs or of reeds has several uses other than as containers: for trays, hats for protection against the sun's heat and armlets or girdles for personal adornment.

The New Stone Age women of the region from which the earliest reliably dated remains of earthenware have come used different quality clays in different localities.

4. Contemporary basket making

By trial and error, they discovered that different colours result from baking a sun-dried vessel in a draught and by excluding air from it, as far as possible, by embedding it in the embers of a fire. Indeed, the workmanship of the first potters was not exclusively utilitarian. From early times, they ornamented their vessels by indentation of the soft clay or by painting it with mineral pigments before drying it.

The earliest villagers of the Middle East—those of Jarmo in Iraq and Jericho in Palestine—left no remains of it behind. They used bowls and other utensils of hollowed stone. Had they already learned to make vessels of earthenware, it is almost certain that excavation would have unearthed fragments from the litter around their dwellings. Whether one or other among later Neolithic communities which practised the art of pottery in the

Middle East or in Egypt invented it is uncertain.

The origin of pottery in the New World is also uncertain, though it is possible that late migrants brought it by the Aleutian Islands route to Alaska. There are many other unsolved enigmas about its origin and distribution. One is that the Bushmen of southern Africa, insulated from even the most primitive barter trade, make their own earthenware vessels. The other is the unequal distribution of pot-making in Oceania. Broadly speaking, the peoples of Oceania are of two sorts. Those called Melanesians make pottery and have little skill as navigators. Those called Polynesians do not make pottery and are able to navigate over long distances. Till recent times, they seem to have been the only link between the Melanesians and the West; and the Melanesians almost certainly reached their island paradises long before the Polynesians got there. This suggests that the know-how of making earthenware vessels did not reach all parts of Oceania from the West.

Both in the New World and in the Old, it is possible to date with more confidence than that of pottery the origin of textile making. Once villagers had mastered the art of making baskets from interlacing twigs, or making mats with reeds and leaves of palm or bamboo, it was not a far cry to weaving; but they had first to learn the trick of spinning. Though the spinning wheel exhibited in antique shops reminds us that spinning was still a household craft in the days of the Pilgrim Fathers, not all of us have a clear notion of what we mean by the process. So a few words of explanation may not be amiss.

If we twist and stretch fibres, it is possible to continue to entangle them into a cord or thread of ever-increasing length and uniform thickness. Provided we twist them tight enough they do not disentangle. Preceding ones

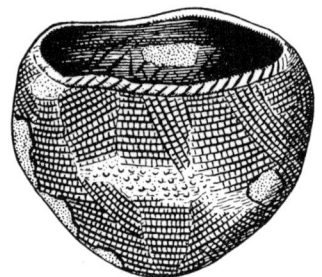

5. Pottery jar from Kenya

help to trap the next ones added. Thus the fibres successively added do not come apart. The thickness of the end product depends on the amount of stretching applied and the rate at which the spinner—man, caterpillar, spider or machine—supplies new strands of fibre.

Amerindians had practised the arts of spinning and weaving several thousand years before Europeans settled in the New World. A technique for cotton spinning still practised in their Guyanese settlements is doubtless akin to the earliest method of spinning. To twist the fine fibres into a continuous thread, the spinner, always a woman, attaches a thick strand to a heavy stone with a hole in it and jerks the stone till it rotates. The inertia of the stone keeps it turning while the free hand adds

6. Neolithic Egyptian pottery

more and more tufts to entangle in the stretched and twisting cord.

Though one usually means by spinning the production of thin threads for needlework and weaving, the process of turning short fibres into a continuous thread is essentially the same as that involved in making string or rope from hemp. At least one cave painting suggests that some of the first of our species knew how to make rope ladders from twisted strands of hay. If they did, we have one clue to what led their descendants to spin thread thin enough for weaving. There is another possibility. Predecessors of the first makers of spun fabrics knew how to make fish nets by twining plant fibres. One such net has turned up in Finland, another in Estonia. Its date, about 6000 B.C., is earlier than that of any remains of fabrics so far found.

Plant fibres spun for weaving include flax, hemp, jute and cotton. Animal fibres include only wool and silk. Of all these, cotton is most interesting because wild species of the plant, whose seed pods supply it as a woolly excrescence, exist in both America and Asia. Botanists place them in the same family as the hibiscus and the hollyhock. The fleece of the sheep supplies the wool of the Old World. Before Columbus, the only source of wool used in America was the fur of three small South American species of the camel family: alpacas, llamas and vicuñas.

Though essentially like making a basket or mat from interlacing twigs or reeds, making a fabric from spun fibre calls for more equipment. At a primitive level this consists of a wooden frame (loom) on which to pin or tie parallel threads (warp) running in one direction, and a bone needle (shuttle) to draw other threads (weft) in the direction at right angles to them. To promote speed of passage for the weft alternately above and below adjacent threads of a warp, the height of alternate threads of the latter should leave a gap just wide enough for the shuttle.

To date, the oldest remains of woven fabric found by excavation of village sites in the Old World are of linen made from flax fibre. They have been found in Egypt at a site dated around 4000 B.C. That flax was the fibre may be merely because it is more durable than wool cloth or cotton fabric. Though we know that the villagers of the Old World cultivated flax at an early date, they probably did so at first as a source, like olives, of vegetable oil. To get fibre to spin from flax, as also from hemp and jute, it is necessary to extract it from the thin stems of the plant by crushing. It is therefore unlikely that flax was the earliest spun fibre, unless some folk of the Old Stone Age had already learned to make stout cords from the crushed stalks with no intention of weaving a cloth.

No preliminary treatment comparable with the crush-

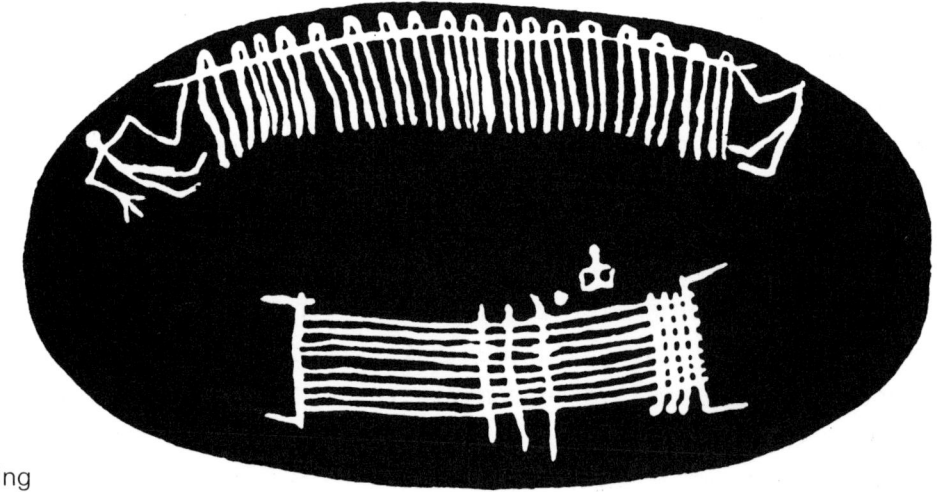

7. Primitive textile making

Handicrafts among the First Farmers 87

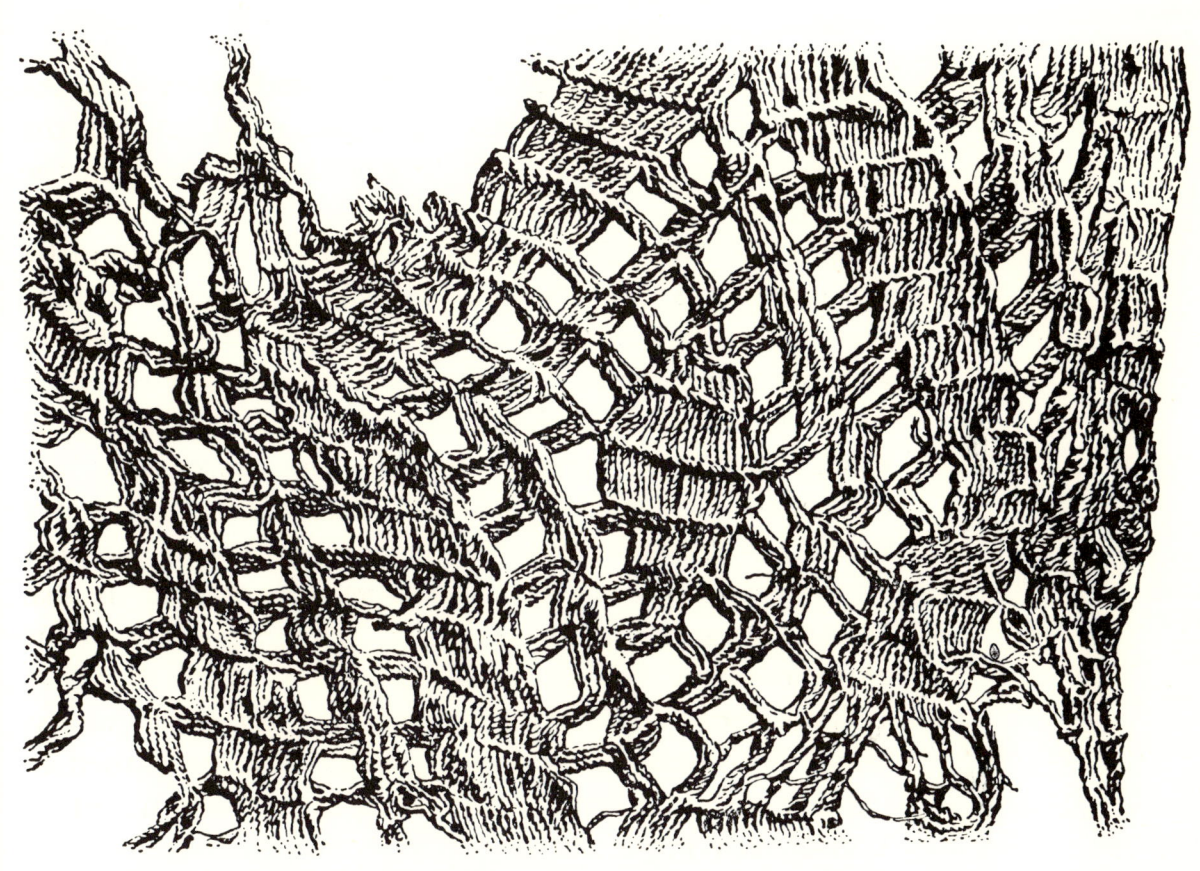

8. Early textile of Peru

ing of flax, hemp or jute stalks is necessary if the fibre for spinning is cotton, wool or silk. For making a fabric, cotton came into use both in India and in America well before 2000 B.C. The use of wool for spinning may have suggested that of cotton in the Old World; but it is unlikely that available sheep had fleeces suitable for shearing, when the womenfolk of the farms in the Old World first wove fabric from spun fibre. There is little doubt that the use of cotton for spinning began long before that of alpaca wool in the New World.

It is tantalising that digging into the past with pick and shovel can rarely uncover relics unable to withstand decay and destruction by weather. Thus common sense tells us that we are as little likely to find early testimony to the practice of some skills as to find fossil jellyfish embedded in rock. One such skill is the manufacture of *bark cloth*, which, though widespread in the hotter parts of both hemispheres, has left no trace in early sites excavated hitherto.

Like leather made from animal hides, bark cloth is not a fabric. To make it, the craftsman, or more usually craftswoman, peels off the inner layers of the bark of a suitable tree, soaks it in water, beats it with a club to the thickness required and then dries it out. We find it used as material for clothing from the Congo across tropical Africa to Madagascar, through Indonesia across the Pacific to Easter Island. Its quality reaches its highest level among the Polynesians and the inhabitants of tropical America. Often the makers adorn it with lively patterns.

Some may see in this confirmatory evidence for the belief that Easter Island has at one time been a half-way house for the diffusion of household skills to the New World from the Old. Others may conclude that men and women like ourselves have come up with the same answers to the same questions when faced with the same needs and the same difficulties. We shall meet this dilemma again when we compare the cradles of civilization in the Old World with the temple cities which existed in America before Europeans arrived as conquerors. Some readers may therefore like to know in bare outline what we can say with confidence about the relations between peoples of the two continents before Leif Ericson (A.D. 1000) skirted the east coast of Canada or New England.

The first facts to bear in mind are geographical. We have referred briefly to one of them in our last chapter. South of the Bering Strait, a gap of less than sixty miles separating Alaska from Siberia, there is a long chain of

9. Bark cloth

islands spanning the northern extremity of the Pacific. The distance between any adjacent pair of the group, called the Aleutian Islands, is not formidable (frequently less than twenty miles and almost never more than a hundred) even in terms of a very primitive level of navigation. South of the Equator through Indonesia, the long chain of islands commonly called Oceania extends east of the Asiatic continent. Among those inhabited before European settlement in the New World, the one nearest the coast of South America is Easter Island, now owned by Chile. It lies 2000 miles west of the Chilean coast and not much less from the nearest Oceanic islands inhabited before European traders visited them.

The southern chain, of which Easter Island is the most eastern outpost, contains many monuments of stone. Among the most impressive of these are those on Easter Island itself: and there is little doubt that its first occupants were Asiatics very skilled in long distance voyages, as are the Polynesians of today. If such Asiatic colonizers could reach Easter Island, they, or their offspring, may well have been capable of covering the distance between Easter Island and Chile.

So much for geography. Let us now take a look at the human relations between the two hemispheres. The stock from which human beings sprung unquestionably evolved in the Old World, and almost certainly within the tropical zone of it. The earliest human remains so far found in America are those of *Homo sapiens,* and the first of our species to reach it did so long before they made boats as good as those of contemporary Polynesians. The first traces of their arrival are indisputably man-made flaked flint knives and spearheads, along with bones of species, such as the mammoth, long since extinct on the American continent. These occur in North America and they belong to a period at least 10,000 years ago. Thus the earliest human inhabitants of America can have come there only by the Aleutian route, and did so before, or about the time, when village life began in the Asiatic Middle East.

If later waves of immigrants arrived by the Easter Island route with some knowledge of cultivation and village crafts, we should expect to find the earliest traces of farming on the west coast of South America. Though further excavation may confirm this, remains found so far do not. The earliest found so far are from a Mexican cave dated as about 6500 B.C. Traces of beans in it indicate that cultivation of legumes began in Mexico before that of maize. The same is true of village sites on the coast of Peru, dated much later, i.e. about 3000 B.C. The people of this region also cultivated beans but no maize. Besides beans, they cultivated squashes and cotton, used crudely flaked cutting tools, and made baskets, mats and rude cloth from coarsely twined cotton fibre. Though they made baskets and mats, they did not make pottery. They lived settled lives, and had underground dwellings. It may be important to record that they practised fishing, being seagoing folk.

In the light of available evidence, it thus seems that the beginnings of village life in South America are later than those in the northern half of the New World. This supports the view that cultural progress spread southward, like the practice of pottery, along the route which the first arrivals followed. So far, therefore, the balance of evidence is against the view that Polynesian navigators brought the knowledge of agriculture to America.

If they ever reached the western hemisphere *via* Easter Island, they must have done so at a date much later than that at which New World farming began in their nearest port of call. Since species suitable for cultivation in the Old World were not to hand in the New, it is difficult to believe that they could have transmitted information of value, even if they had arrived in America before cultivation of crops began. Nor is it likely that waves of Asiatic immigrants who followed the southward trek from Alaska contributed to what some writers call the New World Neolithic Revolution. It is most unlikely that any knowledge of cultivation had reached the Aleutian chain by the time crop production had begun in Mexico. In short, the first settlers to grow crops in America seem to have learned to do so with no assistance from the Old World.

8 Cradles of civilization

Millennia after the so-called Neolithic Revolution in the Old World went by before systematic use of the excrement of herds to renew the soil made more intensive cultivation possible. In most localities, exhaustion of its essential mineral contents therefore set a limit to how many families could congregate on one village site. It also set a limit to how much division of labour was realizable for pursuit of crafts not directly connected with food production. The only exception to this otherwise inexorable rule in the Eastern half of the world was where great rivers annually flooded the territory nearby their banks. There they inundated it perennially and thereby enriched the earth with silt from their sources. This state of affairs existed along the Nile in Africa and along the Tigris and Euphrates in Asia.

As we have seen, the story of civilization in the most literal sense of the term, that is town life, begins there before 3000 B.C. Prosperous riverside village communities coalesced into cities surrounded by satellite farming settlements. Somewhat later, perhaps a thousand years, great concentrations of population also appeared in the watersheds of the Indus of Pakistan and the Yellow River of North China. As yet, we know far less about their beginnings and of how much they owe to the diffusion of new skills along the trade routes which connected them with what we may provisionally call the cradles of civilization in the Old World.

Onwards from about 3500 B.C., such hitherto incomparable concentration of human populations in Egypt and Iraq betoken far more elbow room for division of labour. Hitherto there had been housecrafts for the womenfolk and less skilled labour for the males. Henceforth there were craftsmen of several sorts, mostly male. Such specialization brought with it new inequalities of possessions and of social status. Both in the Old World, and much later in the New, we see the comparatively sudden emergence of a priestly ruling caste withdrawn from food production, homecrafts of village life and manual labour of any sort.

Its advent heralds erection of huge buildings dedicated to ritual and reserved for the residence of the governing priesthood or of their royal overlord. Their lay-out was possible only by enlisting experts in new techniques of measurement. Their construction was possible only by recruiting artificers with new techniques for shaping and placing blocks of stone where the planners dictated the lay-out. Behind the planner and the artificer, there was a labour force of slaves, possibly recruited at first from peasants unable to pay taxes in grain, olives, wine and the like to their priestly overlords.

Rapid progress of technical prowess by the craftsmen who enjoyed priestly patronage could not have occurred before metal tools supplanted the best blades of finely polished stone. The metal first used for tool-making was copper. When copper came into the picture, carpentry and all the activities which depend on wood as raw material took a great leap forward. When city life began in the Middle East, it was already in use in some of the more prosperous villages. In the New World, copper and its alloy bronze were the only metal substitutes for stone before European conquerors came.

The use of copper among the first mixed farmers of the Old World gave their way of life a new look in more ways than one. Traders on overland routes now had asses or camels to carry their wares, of which metal or metal ore was the most weighty. Demand for it therefore encouraged dispatch of bulky cargoes up and down the great rivers on whose banks the cities lay.

Those of Iraq rise in the heart of Armenia. The neighbouring territory of the Caucasus was one of those where metallurgy gained an early footing in the Old World. The Sinai Peninsula near the delta of the Nile was another.

Long before this, the trader boatman and the pedlar with a pack animal had a market for wares with no immediate pay-off in what was to become the world's work. In the more prosperous villages there was a growing demand for mineral pigments, i.e. metallic compounds such as ochre (iron oxide) and cinnabar (mercuric sulphide), to tint patterns on the glaze of pottery. There was also a growing demand for articles of personal adornment. Two of these have a tale to tell about how copper eventually replaced stone for cutting blades.

Before the dawn of city life in Egypt and in the Middle East, there was a market for gold pins, rings, etc. in the more prosperous villages which were merging into larger communities. There was also a demand for an attractive green stone called *malachite*. Malachite is an alkaline copper carbonate. It is attractive as an ornamental stone when polished and, as such, suitable for articles of jewellery. It was also used, when powdered, by highborn Egyptian ladies as eye shadow. Village housewives had already done so before there were cities along the Nile. Digging has unearthed remains of palettes on which they mixed the powder.

Before we can understand where malachite comes into the story of copper tools, we must be clear why the pedlar could find customers for gold articles or for gold to make them with. Many centuries before there were cities, gold had a fascination for mankind out of all proportion to its slender usefulness. Part of the attraction may be its glitter in sunlight. Whatever the explanation of its appeal, its sources and properties make it easy to locate and easy to work with. It is soft and does not tarnish. It occurs as pure metal in nature, and is remarkably widespread. Today, its range extends from Alaska to Peru, from Peru to South Africa and from Siberia to Australia. Its main source in prehistory was the gravel of river beds, where its glitter is visible.

Being one of the heaviest metals, it separates readily by shaking from a watery suspension of the gravel where one finds it. Mostly the particles which separate are tiny; but it is easy to hammer a handful of them into sheets which are flexible and easily cut with a sharp flint blade, if sufficiently thin.

Its softness makes gold suitable for fashioning body ornaments or other *objets d'art* which remain as resplendent after five thousand years as when first made. It is indeed far too soft for making a cutting edge. The first traders in gold, prospectors for gold and smiths who could shape it at will, had therefore no foresight of the use of metals for the manufacture of useful tools and formidable weapons. Such use almost certainly began as a blunder when the prospectors came across native copper and mistook it for gold, but, being hard, for gold with a difference.

Like gold, copper occurs as metal both in the Old and New Worlds. To be sure, native lumps of it do not look like gold; but scratching the corroded surface reveals a reddish metallic glitter. On polishing the nugget with grit, the whole surface will have a metallic lustre. To the prospector, with no knowledge of a metal other than gold in the pure state, it would seem to be a reddish variety of the much coveted element. When beaten with a polished stone hammer, it looked still more like it, suitable at least for beating into beads or bracelets.

We have good reason to believe that prospectors who knew no chemistry, and hitherto knew only one metal, proceeded on this assumption. When Europeans reached the New World they found that copper replaced gold objects of personal adornment in regions where there was no gold but abundant copper. While the farming tools of the Old World tillers were still of polished stone, the villagers made pins of native copper, or obtained them from traders.

It is not entirely true to say that gold had no utilitarian value at an early stage in the story of mankind. Where it was abundant in South America, it had been used before

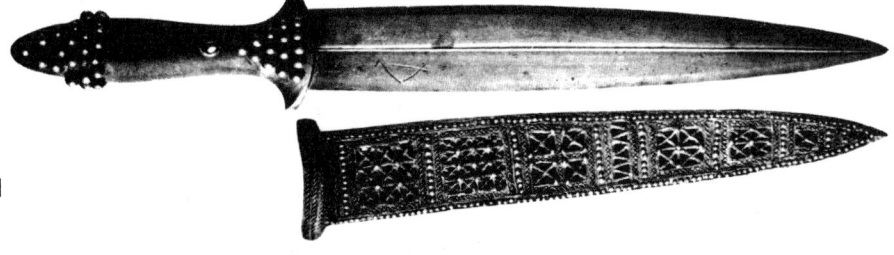

1. Sumerian goldsmith's work
   (a) dagger discovered at Ur (Third Early Dynastic Period)
   (b) gold tumbler and bowl

the Spanish Conquest for making fish hooks. In the New Stone Age villages of the Old World, the earliest remains to date of a useful device made by hammering native copper is a harpoon head. Such a use for copper happened when some of the goldsmiths of the New Stone Age accidentally discovered that native copper hardens when hammered.

Though this makes it less fitting than gold for fashioning pendants, armlets and the like, it does make it suitable for making a tool or weapon. For making tools or weapons, copper therefore gradually displaced flint and such hard stone as obsidian. Archaeologists now speak of a phase which overlaps the beginnings of city life as the *Chalcolithic*. This is Greek for copper-stone. It signifies that there was a span of time when tillers and craftsmen, still using stone hammers and mattocks, were beginning to use copper for cutting tools such as sickles.

To hammer native copper into a serviceable shape was one thing, to extract it from its ore was another. The use of the ore itself was a momentous step, and it led to another. The extraction of the molten metal by people already familiar with making bricks suggested a new method of shaping it. They were not slow to grasp the possibility of using a clay mould to cast it in a suitable configuration. Stone tools lost ground increasingly when craftsmen found out how to do so, and when trade made copper blades widely available. Of all its ores, the one most early mined for Egyptian use was the green carbonate malachite, of which there were deposits in the Sinai Peninsula on the route of traders between Asia Minor and the early city sites of the Nile.

A possible explanation of the beginning of metallurgy in the usual sense of the term, i.e. extraction of metal from its ores (oxides, sulphides, carbonates, silicates, etc) is a household mishap when the powdered malachite, already applied as eye shadow by the women folk of the more prosperous villages, fell into the hearth from the palettes they used for make-up. Spangles of metal with a glitter like that of gold would then appear as the powder mingled with red hot embers of a hotly burning wood fire. An objection to this attractive story is that an open charcoal fire reaches a sufficiently high temperature to bring about the change only if there is a very strong draught.

When cast copper first came into use, a kiln was the only device certainly used for firing earthenware to extract the metal. Any of several copper ores used as coloured glazes to embellish pottery may therefore have supplied the clue to extraction. What is certain is that malachite was, at least in Egypt, the earliest source of copper other than such as exists in the metallic state; that the molten metal is obtainable therefrom by means of a charcoal fire; that it is possible to cast it in any prescribed shape by making a clay mould; and that copper so cast in a mould is of much greater hardness than gold or silver.

The discovery of how to shape it in a mould may well have been made by potters. Thenceforth the emergence of a new body of specialists, the coppersmiths was inevitable. Thereafter, almost every other craft shared in the benefits of a new prospect of greater productivity. The coppersmith could offer better tools to those who felled trees for house building, boat making or bridge construction, to those who trimmed the felled trunk, worked with wood, hoed or dug and ploughed in the fields.

This mounting demand for sources of metal gave the trader prospector and attendant smiths the incentive to venture farther afield especially where there were unexplored waterways with friendly settlements on their banks. In this quest, they had to contend with barriers of mountains and forest, and came to rely more and more on navigable rivers. The search for metals therefore gave a powerful impetus to the construction of larger boats of better design than those which sufficed for the needs of folk who fed themselves largely by fishing.

Every craft which contributes to ship construction shared in the impetus it received from the metal hunger of new concentrations of population with massive building projects. By 3000 B.C. there were boats which could cross the small inland Sea of Azov at the northern tip of the Black Sea or venture into the Mediterranean where there was copper ore to mine in Cyprus. Indeed, the very word copper comes from Latin *cuprum*, itself a corruption of *cyprium*, that is Cyprus-stuff. Before 2500 B.C. seaborne traffic linked Egypt with Crete which became the site of a short-lived and immensely rich civilization till subdued about 1500 B.C. by invasion from the Greek mainland.

After the discovery of how to extract copper and shape it in moulds, the next advance was its substitution for tool and weapon manufacture by its harder alloy *bronze*. Brass is a copper-zinc alloy but bronze consists of copper (about 85 per cent) and tin (about 15 per cent). Probably, the discovery of its usefulness happened before that of a recipe for mixing its ingredients. From Oman on the edge of Saudi Arabia nearest to the mouth of the Persian gulf, the earliest civilization of Iraq imported a copper ore containing up to 14 per cent of tin, and therefore a natural source of the alloy.

In localities where the two existed side by side separately, it was a short step to find that a darker product of greater hardness than pure copper is obtainable by cook-

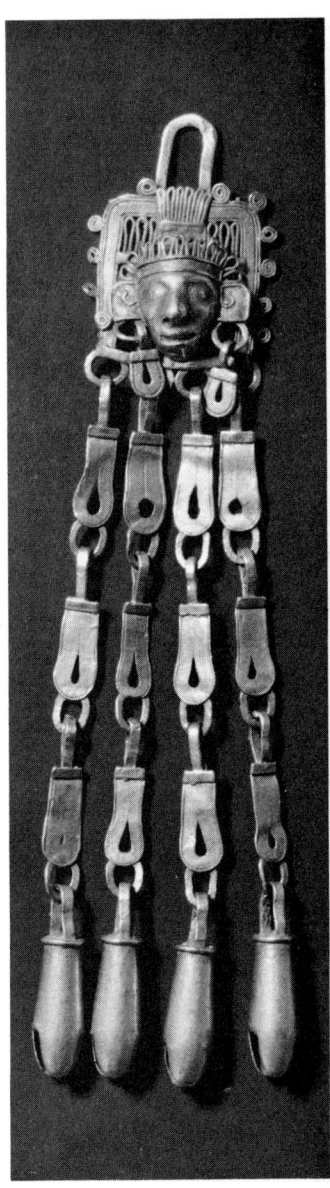

2. Aztec goldsmith's work

ing in the same furnace copper with tin ore. The search for tin ore as a separate ingredient began before 2000 B.C. The earliest source seems to have been the north-east corner of Spain. Thence the search extended east to India and north to Britain. When history first discloses anything substantial about its origin, merchant mariners were already taking long sea voyages to bring back British cargoes of tin to the wealthier communities of the Mediterranean and near East.

The extraction of tin from its ores did not call for more powerful furnaces than does copper, and it is far easier to mould. Whereas the melting point of gold, silver and copper differs little from ten times the boiling point of water, that of tin is less than three times. The first smiths who extracted the pure metal did not clearly distinguish it from lead or zinc whose melting points are both considerably lower than those of copper, gold and silver. Possibly, the pioneers of metallurgy confused tin with silver which gained at an early date in both hemispheres a popularity second only to gold for ornamental use.

Neither zinc nor silver plays any part in the story of how metal replaced stone; and tin is of interest only as a constituent of bronze. The use of bronze for tools and weapons spread farther afield as its sources became more abundant; but the spread was slow. To speak of the Bronze Age can therefore be as misleading as it would be to speak of a Horse Age. In the northern steppes of Asia, bare-back riders who had come to terms with the horse at a much earlier date were still letterless savages with little or no knowledge of agriculture when they came with the first horses into Egypt. From almost every other viewpoint, they were three thousand years or more behind the peoples of Egypt in the arts of farming, building and metal work.

The next milestone of metal use came two thousand years after the first use of bronze in the Old World. While bronze was still a novelty, smiths in different regions had learned to hammer into shape heated fragments of meteorites (thunderbolts) for ornamental beads. Such seems to have been the first use of iron. The beads were possibly marketable as charms because of the awe which the descent of a so-called thunderbolt evoked. They may also have had status value as rarities and curios. The extraction of metal from its ores was inevitably a much later innovation. It did not happen at all in the New World.

That it occurred in the Old World is remarkable. Iron melts only at temperatures seven times as high as the melting point of tin; and bellows sufficiently powerful to produce the necessary blast, whence to cast the metal in a mould, did not come into use before the Christian era. In antiquity the process of extraction and hardening was therefore laborious, and successive stages involved discoveries to which experience gained in working with copper and bronze gave the smiths no clue.

In bare outline, the first ones to work with its ores proceeded as follows. Using a blow-pipe or simple bellows to raise the temperature as high as possible, they roasted one of its native oxides (*haematite* and *magnetite*) in a clay kiln till they obtained a pasty mess called a *bloom*. This breaks up easily on cooling. By hammering and heating repeatedly in a blast of air, it is, however, possible to weld small lumps into a piece of *wrought iron*, which becomes progressively harder by repeatedly hammering it while red hot and suddenly cooling it by immersion in water between successive batterings on the anvil.

The progressive hardening which occurs in this way is due to the production of a surface layer of steel whose thickness depends on repetition. Clearly, therefore, a complete recipe for making wrought iron with a good casing of steel could not have taken shape suddenly. Success came about because the soft iron produced in the initial stages of hammering was already marketable for ornamental purposes centuries before the smiths could foresee the possibility of making a cutting or piercing edge of steel.

Had our ancestors restricted the use of metals to ploughshares, hoes, hammers, cutting tools and domestic amenities, the story of metals would be one of happy achievement. Actually, it is the story of how mankind became more and more destructive and more dependent on the slave labour of war captives.

About 1700 B.C. the Hittites of the part of Asia Minor to the north of Iraq had almost a monopoly of steel in the Middle East and Mediterranean. Their home was a region near to the Caucasus, which was the site of the earliest production of steel. From the Caucasus they acquired the steel weapons with which they overran Iraq. They had another advantage over the regions they conquered. Equipped with horses, from their northern neighbours, the Hittites fought with chariots as well as with steel weapons. Their Syrian neighbours, also equipped with steel, horses and chariots, swept at about the same time into Egypt where they founded the short-lived dynasty of Shepherd kings.

Though the discovery of metals added immeasurably to man's command over nature and scope for inventive ingenuity, material gain played only a minor role in the earliest stages of finding better substitutes for stone tools and weapons. The inclination to beautify pottery and to adorn the human body stimulated a demand for metal and metallic compounds as pigments or as stones for making beads and other ornamental articles. Our account of the cradles of civilization would be incomplete if it made no reference to how expanding trade fostered another medium of artistry.

Among commodities for which the trader boatmen could find customers in the affluent societies of Egypt and Iraq, our list should include musical instruments. Of these the lute was of Northern, possibly Caucasian, origin. In the dawn of city life, priestly ritual and folk dancing enlisted performers with many other types of musical instruments, including clappers and drums, flutes, harps and lyres. Though indisputable relics of none of these come from the excavation of earlier village sites, the variety in use by about 3000 B.C. suggests that instrumental music of some sort was an accompaniment of tribal ritual at a date scarcely less remote than the age of the Cave Painters. It is well-nigh universal among the most backward communities still living.

Instrumental music was probably the most truly popular of the arts in the first of the Old World cities, where craftsmen and cultivators of the soil were becoming less and less privileged as the power of the priestly caste grew and as construction of vast temples and palaces absorbed more of the wealth of the community. Though a rapacious and dictatorial caste, it is wrong to think of the priesthoods of ancient Egypt and Iraq as useless parasites or to speak of the temples as places of worship in the sense of the word current as we use it today. The priests were the custodians of the calendar and the temples were their observatories. Regardless of its relevance to superstitious ceremonial, a calendar is indispensable to a seasonal food economy of which grain

104 Beginnings and Blunders: Before Science Began

3. Ziggurat

MARJORIE V. DUFFELL delt 1937.

growing is the kingpin; and the layout of the so-called temples discloses their use for astronomical observation essential to keeping track of the seasons.

The calendar of the astronomer-priests, the construction of their temple-observatories and the accountancy of the tribute or possessions of the ruling caste encouraged three interconnected intellectual advances of immense influence on the subsequent history of mankind. One was the introduction of signs for making a written record. A second was the discovery of crude recipes for calculation. The third was the discovery of simple rules for measurement. The cultivation of these skills, and therewith the means of transmitting to posterity a lively record of their way of life, places the city communities of Egypt, Iraq and Crete in a category apart from the peoples from whom they learned the craft of metallurgy. In short, it was in such communities that science began.

It seems likely that some of this new knowledge made its way eastwards through India to China, where a comparable priestly caste is a less prominent feature of the pattern of early city life. This possibility leaves unsolved the riddle of how civilization began in the New World. Even if we could be sure that Asiatic immigrants, familiar with the sea lanes of Polynesian navigators, brought to America the know-how for cultivating crops, basket-making and spinning, it is scarcely credible that they could have contributed directly to the native civilizations of Central America (*Maya, Toltec, Aztec*) and of Peru (*Inca*).

Though the earliest of these, the Mayan, came into being only about the beginning of the Christian era, the astronomical knowledge of its priests and the precision of their calendar in some respects surpassed the level of scientific knowledge in the Old World before the time of Hipparchus (150 B.C.). Their division of the calendar, the role of the planet Venus in it and their system of counting in multiples of twenty with a zero sign are each alien to Asiatic custom.

In one way, the location of the pre-Columbian civilizations is very different from that of all the earliest cradles of civilization in the Old World. None of the temple cities of America lies along the banks of the Mississipi or Amazon, its only two great rivers comparable to the Nile, the Tigris and Euphrates, the Indus or the Hwang-ho. There may be a biological explanation of this. Since the Amerindians learned to cultivate beans before they started to grow maize, their soil was less liable to exhaustion of its nitrate content. They were therefore less dependent on periodic renewal with silt from inundation.

Despite all the differences between the cradles of civilization in the Old and the New World, the similarities are impressive. The role of the priesthood as custodians of a ceremonial calendar and the erection of huge temple observatories is common to both. The Aztecs of Mexico built step pyramids which are extraordinarily like those of early Egyptian civilization. Like the desert land around the Great Pyramid, and the Sphinx near Cairo, they now stand on wastes once rich with vegetation.

In both hemispheres, craftsmen practised similar techniques for grinding and polishing stone tools. In both hemispheres, they fashioned gold and silver objects of adornment. In both, they learned to substitute copper and bronze for stone blades. In both, we find the same village crafts; basket-making, bark cloth, pottery, spinning and weaving. Such similarities between the beginnings of civilization in the Old and the New World have given rise to controversies likely to continue till archaeologists unearth far more of our human story.

Though the similarities are striking, a difference between the inventiveness of the cradles of civilization in the two hemispheres is challenging. Before the Spaniards came, people of the New World never made wheels for vehicles. Their failure to do so puts the spotlight on two biological handicaps. There had been wild horses in America before the end of the last Ice Age; but they had become, or were rapidly becoming, extinct by the time the first human beings arrived. When Europeans reached temperate subtropical America, the only herds of hoofed mammals were the bisons, which were too swift of foot and too weighty for corralling before Europeans re-introduced the horse.

Having learned so early to cultivate beans, the Amerindians had less need than the pioneers of Old World village life for flocks and herds. The only beast of burden suitable to their requirements was the llama, one of the small South American slow-moving camels confined to the mountainous part of the tropical belt. The Peruvians used it as a pack animal, like the Old World mule, for transit in the uplands. It was not equal to hard work on the hotter lowlands. Without a beast of burden more swift of foot than the light and leisurely llama, native American civilization had no incentive to construct either a heavy sledge suitable for oxen or a wheeled vehicle suitable for a horse.

When the Spaniards came as conquerors they had all the advantages which helped the Hittites to overrun Iraq three thousand years earlier. They had wheels, they had steel weapons and they had horses. Not one of these was a European, still less a Spanish, contribution to our world-wide civilization. Europeans did not pioneer metallurgy of iron tools and weapons. They did not invent the wheel. Bareback riders had recruited the

4. Mexican step pyramid

horse of the central Asiatic plains as a military ally three thousand years before any European found a use for it other than as meat.

Aside from the biological handicaps which made the conquest a push-over for their invaders, the Mexicans and Peruvians were at an even greater disadvantage than the people vanquished by the Hittites. The European aggressors had cannon; but their gunpowder was not a Spanish invention. It came from China as a harmless recipe for fireworks and became a scourge only when Europeans started to use it for mutual murder in the name of king, country and Christendom.

Thus destruction of the civilizations of Mexico and Peru was not due to inherent superiority of their conquerors. It owed much to the exploitation of discoveries which had aided mankind's search for gold and silver and lust to kill or enslave adversaries considered inferior as heathens.

Chapter I stated that this book is about a kind of

history, called archaeology, through which we can learn about our earliest ancestors, how they hunted and fished, learned to till the soil and produce crops, to construct dwellings and trade between groups along the banks of rivers. Because the surplus of a product was exchangeable by one community for something which was more plentiful elsewhere, improvements of hunting and agricultural techniques, methods of boatbuilding, village construction, weaving, and so forth came about through comparison with the ways of others. Our common civilization is thus the outcome of pooling contributions from people of different continents, beliefs and colour. As human beings, we constantly look forward, and rightly so, to greater achievements, further discoveries. Only by doing so, shall we learn more about our environment. Only by looking back, as we have done in this book, can we learn all there is to know about ourselves.

j573.2
Hogben, L
   Beginnings and blunders

MONTPELIER REGIONAL LIBRARY
RFD #2
MONTPELIER, VT. 05602

Date Due

BTX JUN 1972				
CS MAR 1973	FLX JAN 1976			
MAY 29				
MAY 24		VS DEC 1978		
RSX JUL 1974	CHY JUL 1980			
	FLX JAN	5-13		
	FLX FEB			
74				

VERMONT DEPT. OF LIBRARIES
0 00 01 0287401 3

CAT. NO. 23 231